Single and Multiple Objective

Optimization

All questions and comments concerning this publication should be directed to publisher@weatherfordpress.com.

ISBN-13: 978-1-61580-014-8
ISBN-10: 1-61580-014-X

This book is dedicated to my father who patiently taught me everything about math, physics, and astronomy over countless breakfasts and starry evenings.

Special thanks to Carolyn Jenkins for the cover art.

Table of Contents

1 Single Objective Optimization

1.1 Single Independent Variable

A classic optimization problem is the optimization of a function with a single independent variable. Let

$$y = f(x) \qquad\qquad 1.1$$

Here, y is a dependent variable, x is an independent variable, and $f(x)$ is a twice differentiable function of x. Since $f(x)$ is differentiable, the function is also continuous. In this case, the maximum (minimum) values for the function must occur either at points where the slope of the function is zero, or at the boundary of the range for the x-values. If there are no points where the slope of the function is zero, then the function must be everywhere increasing, decreasing, or constant.

If the function is everywhere increasing, then the maximum (minimum) value of the function occurs at the highest (lowest) x-value on the range of interest. If the function is everywhere decreasing, then the maximum (minimum) value occurs at the lowest (highest) x-value on the range of interest. Finally, if the function is constant everywhere, then there is no maximum (minimum) and every x-value is optimum.

These concepts are illustrated in Figure 1. Here, we are restricted to the x range $0 \le x \le 1$. There are four points where the slope of the function is zero: $x = .15, x = .45, x = .75$, and $x = .95$. These are the extrema we find by setting the derivative of the function equal to zero:

$$f'(x) = 0 \qquad\qquad 1.2$$

Let $\bar{x}_i$ represent the values of x where the derivative is zero. These points are called stationary points. We substitute these values into the function to find the value of the function at each of these points. From this, we find that the highest value is at the point $x = .15$, and the lowest is at $x = .45$.

These are candidates for the global maximum and minimum. We must also look at the value of the function at the endpoints of the range. The point $x = 1$ is between these values. However, the point $x = 0$ is less than the minimum at $x = .45$.

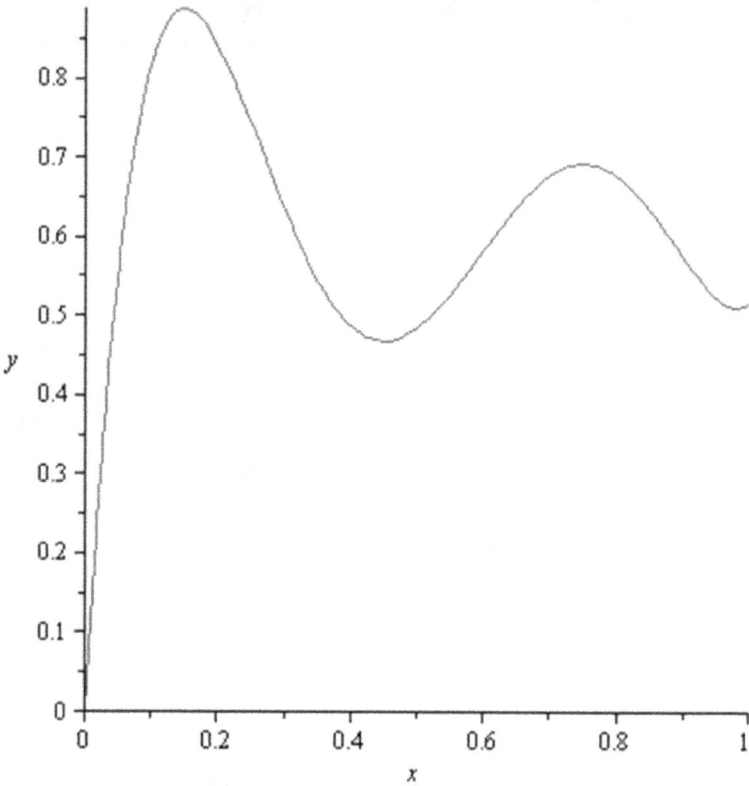

Figure 1: Example function of a single independent variable. There are maxima at the points x=.15 and x=.75. There are minima at the points x=.45 and x=.95. The maximum value on the range is at x=.15, and the minimum is at x=0.

In this case, the global maximum is at $x = .15$, and the minimum is at $x = 0$. We can use the derivative to find the extrema, but we must also examine the endpoints of the range to see if there are values of the function that are higher or lower than the values at the extremum.

Stationary points are not always extremum. Figure 2 provides an example where a stationary point in not in fact an extrema. This function is positively sloped, then flattens out to zero slope, then proceeds again with positive slope. This is called a point of inflection.

If we check the values of the function at the ends of the range, we find that the global maximum occurs at $x = 1$, while the global minimum occurs at $x = 0$. In this case, both the minumim and maximum of the function occur at the boundary of the range for the x-values.

These cases point out the types of issues encountered when examining functions of a single variable. We use the derivative to find the stationary points, but we need to check the boundary to verify these are the global extremes.

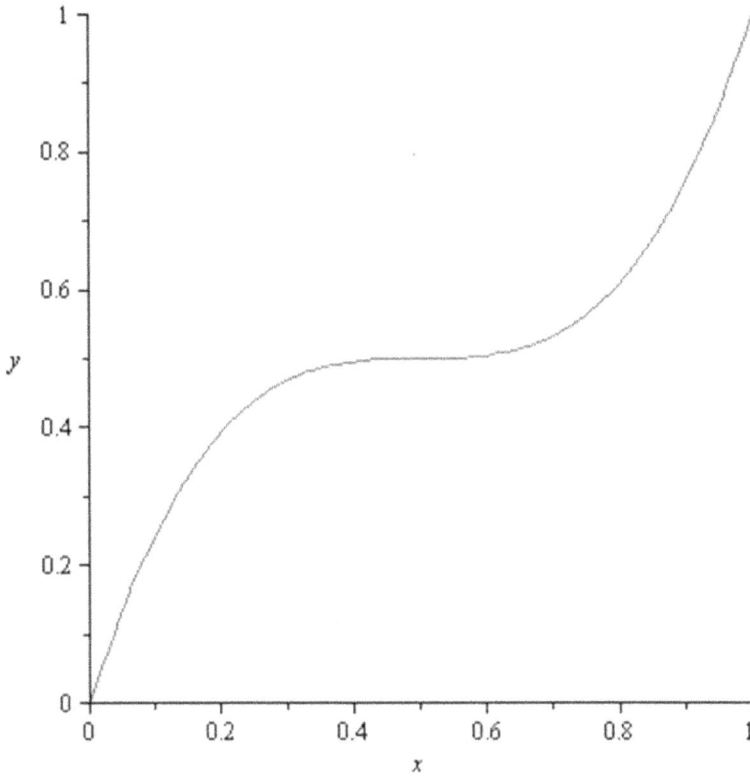

Figure 2: Stationary points are not necessarily minima or maxima. Points of inflection have zero slope, but are neither minima nor maxima.

We may use the second derivative to differentiate stationary points as maxima, minima, or points of inflection:

1. When the second derivative is positive, the stationary point is a minimum.
2. When the second derivative is negative the stationary point is a maximum.
3. When the second derivative is zero, the stationary point is a point of inflection.

The extreme value theorem can be used to determine when a minimum and maximum exist. According to the extreme value theorem, if $f(x)$ is a continuous function on a closed bounded interval $x_{min} \leq x \leq x_{max}$, then there must be values x_{g_min} and x_{g_max} such that

$$f(x_{g\_min}) \leq f(x) \leq f(x_{g\_max}) \ \forall x \in [x_{g\_min}, x_{g\_max}] \qquad 1.3$$

There may be more than one value of x corresponding to the global minimum or global maximum. For example, a function may have a global maximum value of 2, but we might have $f(1) = 2$ and $f(1.2) = 2$. There is only one value for the global maximum, but there is more than one value of x where $f(x)$ has this value.

Using this, our algorithm for determining the extrema of the function is:

Method of Stationary Points I

I-1	Dependent variable y
I-2	Independent variable x
I-3	Differentiable function $f(x)$ relating $y = f(x)$
I-4	Range of allowable values for x: $x_{min} \leq x \leq x_{max}$
S-1	Find the derivative $f'(x)$
S-2	Find $x_i^* = x \mid f'(x) = 0$ where $x_{min} \leq x \leq x_{max}$
S-3	Let N be the total number of points from S-2
S-4	$x_{g\_min} = \min\left(f(x_1^*), f(x_2^*), \dots, f(x_N^*), f(x_{min}), f(x_{max})\right)$
S-5	$x_{g\_max} = \max\left(f(x_1^*), f(x_2^*), \dots, f(x_N^*), f(x_{min}), f(x_{max})\right)$
O-1	Minimum: x_{g_min}
O-2	Maximum: x_{g_max}

Algorithm I: Algorithm for determining the global minimum and maximum for a differentiable function. In the first column, I designates an input, S a step of the algorithm, and O an output.

As an example, examine the function

$$y = \frac{x^3}{3} - 2x^2 + 3x \qquad \qquad 1.4$$

on the range $0 \leq x \leq 5$.

We have all the inputs necessary to use Algorithm I. Executing the first step, we take the derivative of the objective function and set it equal to zero:

S-1	$x^2 - 4x + 3$	1.5
S-2	$x^2 - 4x + 3 = 0$	1.6
S-2	$x_1^* = 1 \quad x_2^* = 3$	1.7
S-3	$N = 2$	1.8

Next, substitute these values back into the objective function to find the value of the objective function at the stationary points:

$$f(x_1^*) = \frac{4}{3} \quad f(x_2^*) = 0 \qquad \text{1.9}$$

We also check the values at the boundaries

$$f(0) = 0 \quad f(5) = \frac{20}{3} \qquad \text{1.10}$$

Putting this together,

S-4
$$x_{g\_min} = min\big(f(x_1^*), f(x_2^*), f(x_{min}), f(x_{max})\big) = 0 \qquad \text{1.11}$$

S-5
$$x_{g\_max} = max\big(f(x_1^*), f(x_2^*), f(x_{min}), f(x_{max})\big) = \frac{20}{3} \qquad \text{1.12}$$

The maximum on the range is $\frac{20}{3}$ and occurs on the boundary at $x = 5$. The minimum is 0 and occurs at the points $x = 0$ (boundary) and $x = 3$.

1.2 Multiple Independent Variables

The previous methods may be extended to the case where there are multiple independent variables. Let

$$z = f(x, y) \qquad \text{1.13}$$

where z is the dependant variable, and x and y are both independant variables. There are multiple dependent variables, but there is still only a single objective to optimize (z).

Stationary points for multivariate functions occur at points where each partial derivative is zero:

$$\frac{\partial z}{\partial x} = 0 \qquad \text{1.14}$$

$$\frac{\partial z}{\partial y} = 0 \qquad \text{1.15}$$

To optimize a function of two variables, we first need a region for the x-y values to search. Designate this region as $\mathcal{R}$. We find all points $(x^*, y^*) \in \mathcal{R}$ where both partial derivatives vanish. Substitute each of these points into the objective function $f(x, y)$, and find the highest and lowest values. Check these values againast the values on the boundary $\partial \mathcal{R}$[1].

Examining the values of the objective function on the boundary is generally a difficult problem. However, when the boundary can be written in a functional form such as

$$y = g(x) \qquad\qquad 1.16$$

we can substitute this form into the objective function, then treat the result as function of a single variable and optimize.

This method can be extended to more than two variables. In general, let the vector $\vec{x}$ represent the independent variables. We take the partial derivative of the objective function with respect to each variable, set each to zero, find all points within the region where all partials vanish, then check against the boundary. These steps are provided as Algorithm II.

Method of Stationary Points II

I-1   Dependent variable z

I-2   Independent variables $\vec{x}$

I-3   Differentiable function $f(\vec{x})$ relating $z = f(\vec{x})$

I-4   Region $\mathcal{R}$ over $\vec{x}$ to find optima

S-1   Find the partial derivatives $\dfrac{\partial f(\vec{x})}{\partial x_k}$

S-2   Find $\vec{x}_i^* = \vec{x} \mid \dfrac{\partial f(\vec{x})}{\partial x_k} = 0 \ \exists k$ where $\vec{x} \in \mathcal{R}$

S-3   Let N be the total number of points from S-2

S-4 $\vec{x}_{g\_min} = \min\left(f(\vec{x}_1^*), f(\vec{x}_2^*), ..., f(\vec{x}_N^*), f(\mathcal{R})\right)$

S-5 $\vec{x}_{g\_max} = \max\left(f(\vec{x}_1^*), f(\vec{x}_2^*), ..., f(\vec{x}_N^*), f(\mathcal{R})\right)$

O-1   Minimum: $\vec{x}_{g\_min}$

O-2   Maximum: $\vec{x}_{g\_max}$

Algorithm II: Algorithm for determining the global minimum and maximum for a differentiable function of multiple independent variables.

---

[1] $\partial \mathcal{R}$ here means the boundary of the region $\mathcal{R}$.

As an example, find the extremum for the function

$$z = x^2 + y^2 - \frac{(x^2 + y^2)^2}{2} \qquad \text{1.17}$$

on the range $|x| \leq 1, |y| \leq 1$.

Following the algorithm,

S-1 $\qquad \dfrac{\partial z}{\partial x} = 2x - 2x(x^2 + y^2) \qquad \dfrac{\partial z}{\partial y} = 2y - 2y(x^2 + y^2) \qquad$ 1.18

S-2 $\qquad \begin{aligned} & 2x - 2x(x^2 + y^2) = 0 \rightarrow x^2 = 1 - y^2, x = 0 \\ & 2y - 2y(x^2 + y^2) = 0 \rightarrow x^2 = 1 - y^2, y = 0 \end{aligned} \qquad$ 1.19

S-2 $\qquad (x_1^*, y_1^*) = (0,0) \quad (x^*, y^*) = \left(x, \pm\sqrt{1 - x^2}\right) \qquad$ 1.20

S-3 $\qquad\qquad\qquad N = 1 + \infty \qquad$ 1.21

The partial derivatives are both zero at the point (0,0). In addition, the partial derivatives are also zero everywhere on the circle $x^2 + y^2 = 1$. There are an infinite number of points on the circle.

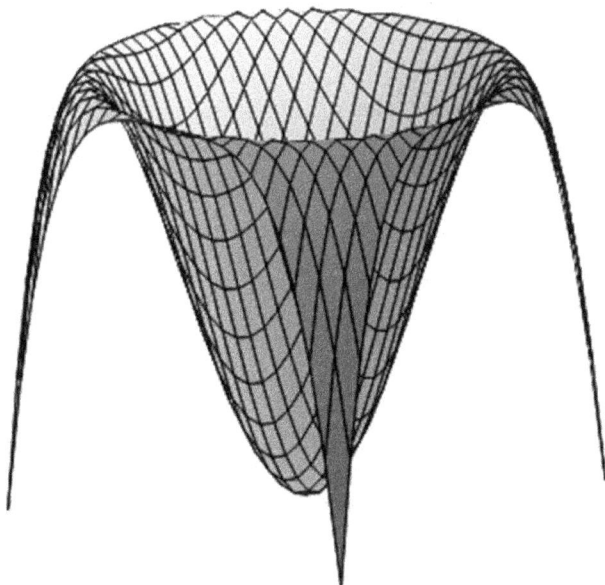

Figure 3: Single objective function with two independent variables.

Figure 3 shows the objective function. The value of the function at the point $(0,0)$ is

$$z = 0 \qquad\qquad 1.22$$

The value of the objective function where $x^2 + y^2 = 1$ is

$$z = 1 - \frac{(1)^2}{2} = \frac{1}{2} \qquad\qquad 1.23$$

We need to examine the value of the function on the boundary. The boundary may be broken up into the four line segments

$$
\begin{aligned}
x &= 1 & -1 \le y \le 1 \\
x &= -1 & -1 \le y \le -1 \\
y &= 1 & -1 \le x \le 1 \\
y &= -1 & -1 \le x \le 1
\end{aligned}
\qquad\qquad 1.24
$$

We examine each of these segments in turn.

$$z(1, y) = 1 + y^2 - \frac{(1 + y^2)^2}{2} \qquad\qquad 1.25$$

Find the stationary points on the boundary. The derivative may be found from equation 1.18 substituting $x = 1$:

$$\left.\frac{\partial z}{\partial y}\right|_{x=1} = 2y - 2y(1 + y^2) = -2y^3 \qquad\qquad 1.26$$

Setting this to zero provides a stationary point at $y = 0$. Thus, we have a stationary point at $(x, y) = (1,0)$. Similarly when $x = -1$, we have a stationary point at $(x, y) = (-1,0)$. Carrying out the same procedure for $y = 1$ and $y = -1$ produces the stationary points $(x, y) = (0,1)$ and $(x, y) = (0, -1)$.

We need to check the value of the objective function at these stationary points, and at the four boundary points $(1,1)$, $(1, -1)$, $(-1,1)$, and $(-1, -1)$. At the stationary points, the objective function has the values

$$
\begin{aligned}
(1,0) \quad & z = \frac{1}{2} \\[4pt]
(-1,0) \quad & z = \frac{1}{2} \\[4pt]
(0,1) \quad & z = \frac{1}{2} \\[4pt]
(0,-1) \quad & z = \frac{1}{2}
\end{aligned}
\qquad\qquad 1.27
$$

On the four boundary intersections the objective function is

$$\begin{array}{lll}(1,1) & z = 0 \\ (-1,1) & z = 0 \\ (1,-1) & z = 0 \\ (-1,-1) & z = 0\end{array} \qquad \text{1.28}$$

Putting this together,

S-4
$$(x,y)_{g\_min} = 0 \qquad \text{1.29}$$

S-5
$$(x,y)_{g\_max} = \frac{1}{2} \qquad \text{1.30}$$

The minimum occurs at the points

S-4
$$(x,y) = (0,0), (1,1), (1,-1), (-1,1), (-1,-1) \qquad \text{1.31}$$

The maximum occurs at

S-5
$$(x,y) = (1,0), (-1,0), (0,1), (0,-1), \left(x, \pm\sqrt{1-x^2}\right) \qquad \text{1.32}$$

1.3 Multivariate Functions

As we examine multivariate functions in later sections, we will make use of some multivariate calculus concepts. These concepts provide insight to the curvature of the objective function at a given point.

One concept is the gradient of a function. The gradient takes a scalar function and creates a vector-valued function based on the partial derivatives:

Gradient
$$\nabla f(\vec{x}) = \begin{pmatrix} \dfrac{\partial f(\vec{x})}{\partial x_1} \\ \dfrac{\partial f(\vec{x})}{\partial x_2} \\ \vdots \\ \dfrac{\partial f(\vec{x})}{\partial x_k} \end{pmatrix} \qquad \text{1.33}$$

The gradient for multivariate functions is in some ways analogous to the derivative for a single variable function.

The Hessian is analogous to the second derivative. The Hessian is the matrix formed from all second order partial derivatives of the function:

| Hessian | $$Hf(\vec{x}) = \begin{bmatrix} \dfrac{\partial^2 f}{\partial x_1{}^2} & \cdots & \dfrac{\partial^2 f}{\partial x_1 \partial x_k} \\ \vdots & \ddots & \vdots \\ \dfrac{\partial^2 f}{\partial x_k \partial x_1} & \cdots & \dfrac{\partial^2 f}{\partial x_k{}^2} \end{bmatrix}$$ | 1.34 |

1.4 Constrained Optimization

The previous sections examined optimization of single objective functions constrained to a specific region of the independent variables. More generally, there may exist more complicated constraint functions that limit the choice of values for the independent variables.

The classic methods for attaching the single objective constrained optimization problem is the method of substitution and Lagrange multipliers. The method of substitution has limited applicability, but commonly appears in practice. The method of Lagrange multipliers is more generally applicable, but requires additional mathematical machinery.

We desire to find the optimum for a differentiable function on many variables

$$y = f(\vec{x}) \tag{1.35}$$

subject to the constraint

$$g(\vec{x}) = c \tag{1.36}$$

where c is a constant.

As a concrete example, suppose we want to optimize the function

$$y = x_1^2 + x_2^2 - x_2^2 x_1^2 - 2x_1 \tag{1.37}$$

subject to the constraint

$$x_1^2 + x_2^2 = 1 \tag{1.38}$$

The objective function is a hyperboloid and the constraint function is a cylinder. These functions are shown in in Figure 4. The optimization problem amounts to finding the highest (lowest) value of the objective function where the objective and constraint functions intersect.

We solve this using the method of substitution. First we solve the constraint function for x_2^2:

$$x_2^2 = 1 - x_1^2 \qquad \text{1.39}$$

Next we substitute the value for x_2^2 into the objective function:

$$y = x_1^2 + (1 - x_1^2) - (1 - x_1^2)x_1^2 - 2x_1 \qquad \text{1.40}$$

This function only has a single variable. We can identify the stationary points by setting the derivative of the function to zero. By using the substitution method, we have transformed the original 2-dimensional problem into a 1-dimensional optimization problem. The 1-dimensional optimization problem does not require computation of partial derivatives or simultaneously setting two equations to zero to find the stationary points.

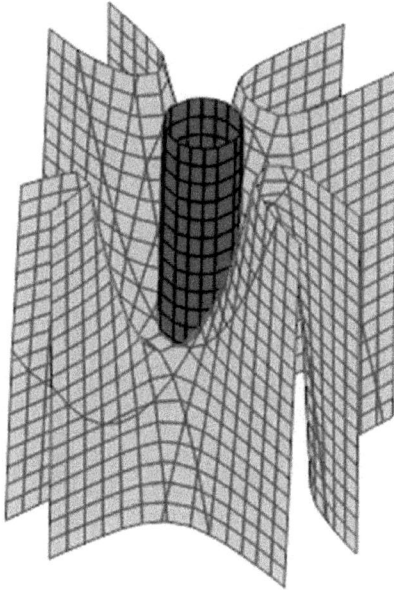

Figure 4: Diagram for the example constrained optimization function. The objective function is the hyperboloid while the constraint is the cylinder.

First, simplify the objective function

$$y = 1 - x_1^2 + x_1^4 - 2x_1 \qquad \text{1.41}$$

Setting the derivative to zero,

$$-2x_1 + 4x_1^3 - 2 = 0 \qquad \text{1.42}$$

Factoring,

$$(x_1 + 1)\left(x_1 - \left(\frac{7}{4} + \frac{1}{2}i\right)\right)\left(x_1 - \left(\frac{7}{4} - \frac{1}{2}i\right)\right) = 0 \qquad \text{1.43}$$

There is one real solution and two complex solutions. The complex solutions are unobtainable because the variables must be real valued. This leaves a stationary point at $x_1 = -1$. The value of the objective function is found by substituting this value into equation 1.43. The objective function is $y = 3$.

The constraint bounds x_1 to the range $-1 \leq x_1 \leq 1$. Similarly, $-1 \leq x_2 \leq 1$. The stationary point at $x_1 = -1$ is at one end of the range. We need to check the other end of the range. Substituting $x_1 = 1$ into equation 1.43, we find $y = -1$.

Figure 5 presents the curve representing the intersection between the objective function and the constraint. We see in these diagrams that the intersection curve is a closed curve with a maximum near $x_1 = -1$ and a minimum near $x_1 = 1$. The minimum is $y = -1$ at $x_1 = 1$, and the maximum is $y = 3$ at $x_1 = -1$.

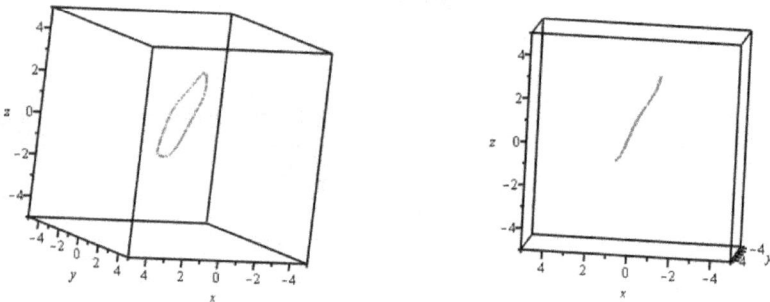

Figure 5: Two views of the intersection between the objective function and constraint from Figure 4.

We were able to find the extreme values of the objective function by substituting the constraint into the objective function then treating the result as a function of a single variable. This approach is useful when the constraint can be solved for one of the independent variables. However, constraint functions are not always approachable with this technique.

Substitution

I-1   Dependent variable z

I-2   Independent variables $\vec{x}$

I-3   Differentiable function $f(\vec{x})$ relating $z = f(\vec{x})$

I-4   Constraint equation $g(\vec{x}) = 0$

S-1   Solve the constraint equation for one of the variables:

$$x_k = h(x_1, x_2, \ldots, x_{k-1}, x_{k+1}, \ldots)$$

S-2   Use S-1 to eliminate x_k from the objective function

S-3   Solve the resulting objective function using unconstrained techniques

Algorithm III: Substitution method is useful for eliminating constraint equations.

We examine the problem again using the technique of Lagrange multipliers. This technique creates a new objective function by adding the original objective function to the constraint equation using a undetermined factor of λ:

$$L = x_1^2 + x_2^2 - x_2^2 x_1^2 - 2x_1 + \lambda(x_1^2 + x_2^2 - 1) \qquad \text{1.44}$$

The stationary points are found by setting the partial derivatives to zero:

$$\frac{\partial L}{\partial x_1} = 2x_1 - 2x_2^2 x_1 - 2 + 2\lambda x_1 = 0 \qquad \text{1.45}$$

$$\frac{\partial L}{\partial x_2} = 2x_2 - 2x_1^2 x_2 + 2\lambda x_2 = 0 \qquad \text{1.46}$$

$$\frac{\partial L}{\partial \lambda} = x_1^2 + x_2^2 - 1 = 0 \qquad \text{1.47}$$

The final equation is the constraint equation. Essentially, this technique creates a single objective function from the original objective function and the constraint equation. By creating a combined constraint, the technique reproduces the constraint and the constraint mixed with the partial derivatives of the original objective function.

The resulting equations are not easily solvable. Numerical techniques may be required to obtain useful solutions. In this case, we can use numerical techniques to solve these equations. We find the two real-values solutions as $(x, y, \lambda) = (1,0,0)$ and $(x, y, \lambda) = (-1,0,-2)$. These are the same solutions we found using the substitution technique.

Lagrange Multipliers

I-1   Dependent variable z

I-2   Independent variables $\vec{x}$

I-3   Differentiable function $f(\vec{x})$ relating $z = f(\vec{x})$

I-4   Constraint equation $g(\vec{x}) = 0$

S-1   Create a combined objective function as

$$L = f(\vec{x}) + \lambda g(\vec{x})$$

S-2   Compute the partial derivatives of L with respect to each of the variables and λ

S-3   Solve the resulting equations for the variables and λ

O-1 k solutions $(\vec{x}_k, \lambda_k)$

Algorithm IV: The method of Lagrange multipliers is generally useful for solving constrained optimization problems on a single objective function.

One of the most useful aspects of the method of Lagrange is that this method creates an unconstrained optimization problems from a constrained optimization problem. Thus, we can focus on developing techniques to solve unconstrained optimization problems. Once these techniques are developed, if we run across a constrained optimization problem, we simply use the method of Lagrange to convert this to an unconstrained optimization problem, then apply the techniques we have previously developed.

1.5 Single Objective Optimization

Formally, we formulate the problem of single objective optimization as

$$max\ f(\vec{x}) \tag{1.48}$$

$$g_k(\vec{x}) \leq 0 \quad k = 0 \dots N_k \tag{1.49}$$

$$h_l(\vec{x}) = 0 \quad l = 0 \dots N_l \tag{1.50}$$

We desire to maximize the objective function $f(\vec{x})$ subject to N_k inequality constraints $g_k(\vec{x})$ and N_l equality constraints $h_k(\vec{x})$. The result is a set of points $\vec{x}_n$ where the objective function obtains its maximum and minimum values while at the same time satisfying the constraints.

Optimization problems are often quite difficult to solve. In the previous sections, we demonstrated the classical techniques for finding solutions to single objective optimization problems.

The method of stationary points is useful for examining unconstrained single objective optimization problems. Alternatively, the methods of substitution and Lagrange multipliers is useful for solving constrained single objective optimization problems.

However, all of these techniques require that the objective function is differentiable. This assumption has a wide range of applicability, but many interesting problems are either not differentiable or not easily differentiable.

For example, problems may have an objective function with N different variables where evaluating the objective function at a single point might require a complicated numerical computer simulation. The objective function might in fact be differentiable, however, since the function is not expressed in terms of elementary functions, there is no method to symbolically differentiate it.

We could attempt numerical integration, but when there are N different variables, we need to evaluate the objective function $N + 1$ times in order to find a numerical derivative. All of these evaluations need to be 'close' together in order for the numerical differentiation to be useful. But this function may not vary quickly enough for a useful difference to arise, or there may be a statistical error associated with evaluation of the function that could render the numerical differentiation useless. Alternatively, the objective function may be quite complicated, and the time required to conduct $N + 1$ evaluations to find a derivative at a single point prohibitively expensive.

For any of these reasons, we need to develop techniques that do not rely on the differentiation of the objective function. There are many such techniques, each useful under certain conditions and less desirable under others. In the next chapter we review several popular techniques for finding solutions to single objective optimization problems.

2 Single Dependent Optimization Techniques

2.1 Preliminaries

In this chapter we review several single objective optimization techniques. These techniques are treated only briefly here to demonstrate the general character of the method. Each of these techniques is deserving of an entire book to show the detailed nuances and capabilities.

Although these techniques are quite different in approach, the underlying problems are the same. Because of this, the techniques have similar inputs and outputs. In this section we examine these common attributes.

The desire is to maximize or minimize an objective function. The objective function is a function of some set of variables that form the input space for the problem. Given a particular set of input variables, we may use the objective function to create a single valued number. It is this number we wish to maximize or minimize.

If there are constraint equations present, we may use the technique of Lagrange multipliers to change the constrained optimization problem into an unconstrained problem. We need to use one Lagrange multiplier for every constraint equation. Essentially, this increases the dimension of the input space in exchange for transforming the constrained optimization problem with an unconstrained optimization problem.

Let $\vec{x}$ be the variables for the input space, and let $f(\vec{x})$ be the objective function. A solution is a particular value for $\vec{x}$ that maximizes (minimizes) the objective function. There may be more than one such $\vec{x}$ because the objective function may have multiple points where it achieves the same maximum (minimum) value.

The maximum (minimum) value for the objective is called the global maximum (minimum). Typically we are interested in finding this value. However, in some cases we may be interested in finding 'local' maxima (minima). Local maxima (minima) are values of $\vec{x}$ where the objective function has zero slope (all partial derivatives are zero). Some algorithms are designed to find local maxima (minima) as well as the global maxima (minima).

We use these terms throughout this chapter as we discuss the techniques for single objective optimization. Many of these techniques are 'evolutionary' algorithms that find maxima (minima) through a series of repeated evaluations of the objective function.

2.2 Examination of Stationary Points

Examination of stationary points was treated in sections 1.1 and 1.2. This approach differentiates the objective function, and sets each partial derivative to zero. Every solution to these equations is a stationary point of the objective function. The objective function is evaluated at these points, and the largest and smallest values of the objective function are recorded.

In addition, the boundary of the region is also checked with the objective function. This result is compared with the largest and smallest values from the stationary points. The overall largest and smallest values of the objective function based on the stationary points and the value of the objective function at the boundary identifies the global maximum and minimum.

The advantage of this method is that the global maxima and minima may be identified through calculus and exact solutions are found. When this method is successful, exact solutions for the global maxima/minima are identified. Also, the local maxima and minima may be found from the stationary points.

The difficulty is this method may lead to systems of equations that cannot be solved exactly in closed form. Numerical techniques may be required to find solutions. Moreover, this method is limited to differentiable objective functions.

Method of Stationary Points II

I-1	Dependent variable z
I-2	Independent variables $\vec{x}$
I-3	Differentiable function $f(\vec{x})$ relating $z = f(\vec{x})$
I-4	Region $\mathcal{R}$ over $\vec{x}$ to find optima
S-1	Find the partial derivatives $\dfrac{\partial f(\vec{x})}{\partial x_k}$
S-2	Find $\vec{x}_i^* = \vec{x} \vert \dfrac{\partial f(\vec{x})}{\partial x_k} = 0 \; \exists k$ where $\vec{x} \in \mathcal{R}$
S-3	Let N be the total number of points from S-2
S-4	$\vec{x}_{g\_min} = \min\bigl(f(\vec{x}_1^*), f(\vec{x}_2^*), \dots, f(\vec{x}_N^*), f(\mathcal{R})\bigr)$
S-5	$\vec{x}_{g\_max} = \max\bigl(f(\vec{x}_1^*), f(\vec{x}_2^*), \dots, f(\vec{x}_N^*), f(\mathcal{R})\bigr)$
O-1	Minimum: $\vec{x}_{g\_min}$
O-2	Maximum: $\vec{x}_{g\_max}$

Algorithm V: Algorithm for the method of stationary points.

Examples of stationary points in one and two dimensions

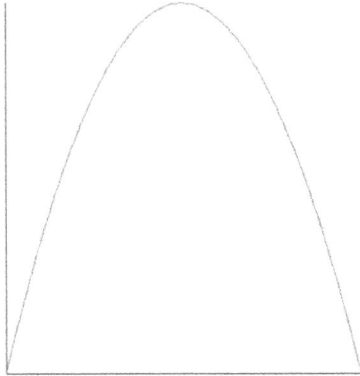

Figure 6: Maximum in a 1-dimensional function.

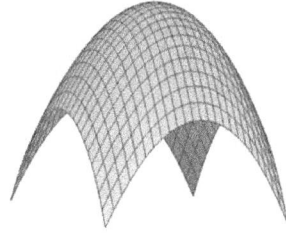

Figure 7: Maximum in a 2-dimensional function.

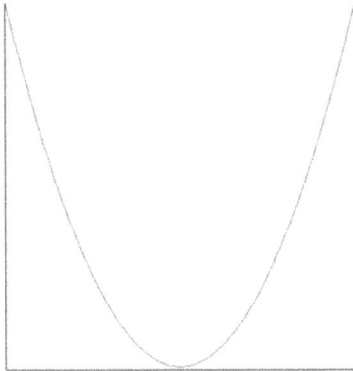

Figure 8: Minimum in a 1-dimensional function.

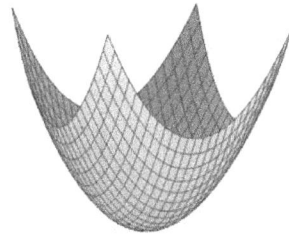

Figure 9: Minimum in a 2-dimensional function.

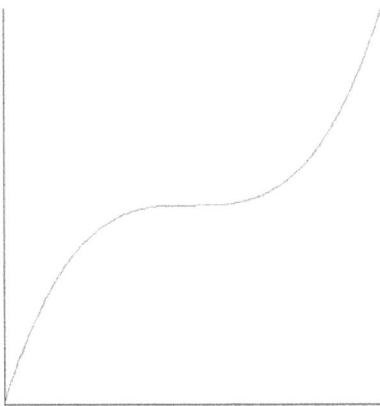

Figure 10: Point of inflection in a 1-dimensional function.

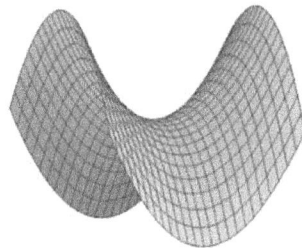

Figure 11: Saddle point in a 2-dimensional function.

2.3 Method of Substitution

The method of substitution is used to eliminate constraint equations. Constraint equations are solved in terms of one of the independent variables, and this is substituted into the original objective function.

The method of substitution was examined in section 1.3. If all constraint equations can be eliminated using this method, then the resulting objective function may be optimized using unconstrained optimization methods.

However, even if only some of the constraint equations are eliminated, optimization of the objective function is more easily approached with fewer constraint equations. The remaining constraints may be eliminated via the method of Lagrange multipliers (see § 2.5).

The main disadvantage is that this method is not always applicable. For example, constraint equations such as

$$x^2 + \ln x = y + \ln y \qquad\qquad 2.1$$

cannot be solved for either variable. In some cases the method may still be applied if the objective function takes in particular forms. Continuing with the example, if the objective function is

$$f(x, y) = (y + \ln y)^2 + x^2 \qquad\qquad 2.2$$

we may still use the method of substitution to arrive at the single variable objective function

$$f(x, y) = (x^2 + \ln x)^2 + x^2 \qquad\qquad 2.3$$

Substitution

I-1   Dependent variable z

I-2   Independent variables $\vec{x}$

I-3   Differentiable function $f(\vec{x})$ relating $z = f(\vec{x})$

I-4   Constraint equation $g(\vec{x}) = 0$

S-1   Solve the constraint equation for one of the variables:
$$x_k = h(x_1, x_2, \dots, x_{k-1}, x_{k+1}, \dots)$$

S-2   Use S-1 to eliminate x_k from the objective function

S-3   Solve the resulting objective function using unconstrained techniques

Algorithm VI: Substitution method is useful for eliminating constraint equations.

Alternatively, there are cases where we can eliminate one variable but not another. For example, consider the objective function

$$f(x,y) = y^2 + x^2 \qquad 2.4$$

with the constraint

$$x^2 + y + \ln y = 2 \qquad 2.5$$

The constraint cannot be solved for y, but can be solved for x:

$$x^2 = 2 - y - \ln y \qquad 2.6$$

This can be substituted into the objective function

$$f(x,y) = y^2 + 2 - y - \ln y \qquad 2.7$$

This removes x from the objective function and reduces the number of variables by one.

As an additional example, examine the objective function

$$f(x,y) = xz \ln(xz) + y^2 \qquad 2.8$$

subject to the constraint

$$2xz\ln(xz) + 3y = 5 \qquad 2.9$$

One option is to solve the constraint for y and substitute the result into the objective function. Solving the constraint for y,

$$y = \frac{5 - 2xz\ln(xz)}{3} \qquad 2.10$$

Substituting into the objective function,

$$f(x,y) = xz \ln(xz) + \left[\frac{5 - 2xz\ln(xz)}{3}\right]^2 \qquad 2.11$$

Here we have reduced the number of variables in the objective function by one as expected. However, we could solve the objective function for the other term:

$$xz\ln(xz) = \frac{5 - 3y}{2} \qquad 2.12$$

Since this term appears in the objective function, we can eliminate this term. In this case we can eliminate two variables from the objective function:

$$f(x,y) = \frac{5 - 3y}{2} + y^2 \qquad 2.13$$

2.4 Aggregation of Constraints

Many optimization problems have an objective function and one of more constraints. The single objective optimization problem may be construed as

$$min\ F(\vec{x})$$
$$such\ that$$
$$g_1(\vec{x}) = 0$$
$$g_2(\vec{x}) = 0$$
$$\vdots$$
$$g_k(\vec{x}) = 0$$
$$h_1(\vec{x}) \le 0$$
$$h_2(\vec{x}) \le 0$$
$$\vdots$$
$$h_l(\vec{x}) \le 0$$

2.14

In the above expression, we desire to minimize the single-valued objective function $F(\vec{x})$. This is a function that takes a vector as an argument and outputs a single value. The may be expressed as the map $F: \mathbb{R}^n \to \mathbb{R}$.

In addition to minimizing the objective function $F(\vec{x})$, there are also several constraint functions. The constraints are divided into two types: equality constraints and inequality constraints. The equality constraints are expressed as the k functions $g_k(\vec{x})$, each of which must be zero for allowable solutions.

Furthermore, there are l inequality constraints of the form $h_l(\vec{x}) \le 0$. Allowable solutions must meet each of these inequality constraints as well as the equality constraints $g_k(\vec{x})$.

The equality and inequality constraints impose restrictions on the values of $\vec{x}$ that are allowed. In some cases, the constraints may be easily implemented based on the coordinate system. For example, in Cartesian coordinates, the constraint $0 \le x \le 1$ is easy to implement. When generating Cartesian vectors, we simply assure that the x-component is on the allowable range.

However, in many cases, the constraints are not as easy to implement and may require solving complex equations or a system of equations. In these cases it is desirable to have a method for optimizing the objective function while simultaneously satisfying the constraints.

One method for accomplishing this is to incorporate the constraints into the objective function. For example, if we have an objective function $F(\vec{x})$ subject to the constraint $g(\vec{x}) = 0$, we might examine the aggregated constraint

$$L = F(\vec{x}) + \lambda |g(\vec{x})|$$

2.15

where we desire to minimize $F(\vec{x})$. If the constraint function $g(\vec{x})$ deviates from the desired value of zero, the value of the aggregate function L increases. Moreover, by increasing the value of λ, we can control how sensitive the optimization is toward the constraint versus the objective function. When λ is much larger than $F(\vec{x})$, optimization will favor the constraints over the objective function.

Inequality constraints may be handled similarly. If we have an objective function $F(\vec{x})$ subject to the constraint $h(\vec{x}) \leq 0$, examine the aggregated constraint

$$L = F(\vec{x}) + \eta \left| max\left(0, h(\vec{x})\right) \right| \qquad \text{2.16}$$

Here, the aggregated term is zero when the constraint is met, then increases as the constraint strays farther from the desired value. In both cases, it is important to note that the aggregated term increases the further we diverge from the constraint. This allows optimization algorithms to gauge the degree of constraint violation and move in the correct direction.

Aggregation of Constraints

I-1   Dependent variable z

I-2   Independent variables $\vec{x}$

I-3   Differentiable function $f(\vec{x})$ relating $z = f(\vec{x})$

I-4   Equality constraint equations $g_k(\vec{x}) = 0$ where $k = 1 \ldots k_{max}$

I-5   Equality constraint equations $h_l(\vec{x}) \leq 0$ where $l = 1 \ldots l_{max}$

I-6   There must be at least one constraint. Thus,

$$k_{max} + l_{max} > 0$$

I-7   Single objective optimization method

S-1   Create a combined objective function as

$$L = f(\vec{x}) + \sum_{k=1}^{k_{max}} \lambda_k |g_k(\vec{x})| + \sum_{l=1}^{l_{max}} \eta_l \left| max\left(0, h_l(\vec{x})\right) \right|$$

S-2   Apply the single optimization method to the function L defined in step S-1 above

O-1   Solutions from the optimization method

Algorithm VII: The method of aggregation of constraints multipliers is useful for solving constrained optimization problems.

2.5 Lagrange Multipliers

The method of Lagrange multipliers is another method for eliminating constraint equations. Here, we modify the objective function to include additional terms, where each terms is a constraint equation multiplied by an unknown constant (the Lagrange multiplier).

The advantage for this method is that it applies to a wide range of constraint equations. In fact, the method may even be used when the objective function is not differentiable. It is required that the partial derivatives with respect to the Lagrange multipliers exists, but this is satisfied by the form of the objective function determined from the method.

The main disadvantage is that this method increases the dimensionality of the objective function. Each constraint equation is multiplied by an independent unknown constant, and each of these constants adds an additional dimension to the objective function. By increasing the dimensionality of the space for the objective function, we increase the difficulty of finding solutions.

If we start with an initial constrained optimization problem with objective function $f(\vec{x})$ with N constraint equations of the form $g_k(\vec{x}) = 0$, we transform the optimizing the unconstrained objective function

$$f(\vec{x}) + \sum_{k=1}^{N} \lambda_k g_k(\vec{x}) \qquad\qquad 2.17$$

Lagrange Multipliers

I-1	Dependent variable z
I-2	Independent variables $\vec{x}$
I-3	Differentiable function $f(\vec{x})$ relating $z = f(\vec{x})$
I-4	Constraint equation $g(\vec{x}) = 0$
S-1	Create a combined objective function as $$L = f(\vec{x}) + \lambda g(\vec{x})$$
S-2	Compute the partial derivatives of L with respect to variables and λ
S-3	Solve the resulting equations for the variables and λ
O-1	N solutions $(\vec{x}_k, \lambda_k)$

Algorithm VIII: The method of Lagrange multipliers is generally useful for solving constrained optimization problems.

As an example, examine the objective function

$$f(x, y) = x^2 + y^2 \qquad \text{2.18}$$

subject to the constraint

$$xy + e^{xy} = 1 \qquad \text{2.19}$$

The constraint cannot be easily solved for either variable. Here the substitution method does not provide a method to incorporate the constraint into the objective function.

Write the constraint in the form

$$xy + e^{xy} - 1 = 0 \qquad \text{2.20}$$

Applying the method of Lagrange, we create a new objective function as

$$f(x, y) = x^2 + y^2 + \lambda(xy + e^{xy} - 1) \qquad \text{2.21}$$

We minimize this function with respect to the three variables x, y, and λ. Take the partial derivative with respect to each of the variables and set the result to zero:

$$\frac{\partial f}{\partial x} = 2x + \lambda(y + ye^{xy}) = 0 \qquad \text{2.22}$$

$$\frac{\partial f}{\partial y} = 2y + \lambda(y + xe^{xy}) = 0 \qquad \text{2.23}$$

$$\frac{\partial f}{\partial \lambda} = xy + e^{xy} - 1 = 0 \qquad \text{2.24}$$

Minimizing the new objective function with respect to λ becomes the original equation for the constraint. The method of Lagrange incorporated the constraint into the objective function be adding an additional parameter such that when we minimize the new objective function with respect to this parameter, we regain the original constraint equation.

The resulting system of equations must be simultaneously solved for the original variables as well as the Lagrange multipliers. In some cases this may be feasible, but as the simple example above illustrates, in many cases these equations cannot be easily solved.

We can obtain a solution to the simultaneous equations using numerical methods. Alternatively, we may minimize the new higher dimensional objective function using other techniques in this section. In either case, the method of Lagrange converts a constrained optimization problem into an unconstrained optimization problem.

2.6 Newton's Method

Newton's method is a method for numerically computing the zeros of a one dimensional function over a predetermined range. When the objective function is differentiable, we may take the derivative of the objective and set it to zero. Newton's method may be used to determine the zeros of the derivative to numerically find the values of the independent variable that locates the stationary points.

Newton's method begins using some predetermined test point x_0. This point is the starting point for the algorithm. If we have some knowledge of where the stationary point may be, we choose x_0 in the vicinity.

The method will proceed iteratively. We start with an initial point x_0, then compute a 'next guess' x_1, then compute another point x_2, etc. To determine the sequence of points, we expand the objective function as a Taylor series about the current location (x_n)

$$f(x + x_n) = f(x_n) + (x - x_n)f'(x_n) + \frac{(x - x_n)^2}{2}f''(x_n) + \cdots \qquad 2.25$$

At a stationary point, the derivative must be zero. The derivative is

$$f'(x + x_n) = f'(x_n) + (x - x_n)f''(x_n) + \cdots \qquad 2.26$$

If the derivative is zero,

$$f'(x_n) + (x - x_n)f''(x_n) = 0 \qquad 2.27$$

$$x = x_n - \frac{f'(x_n)}{f''(x_n)} \qquad 2.28$$

We use this as the next point in the sequence. Thus,

$$x_{n+1} = x_n - \frac{f'(x_n)}{f''(x_n)} \qquad 2.29$$

Essentially, Newton's method assumes we are near a stationary point, then approximates the local behavior of the function as a quadratic. We solve the quadratic to find the point where the quadratic is expected to have its stationary point. We use this as the next point in the sequence of the algorithm.

This method will quickly converge to the stationary point when we are near the stationary point and when the second derivative is nonzero in the vicinity of the stationary point. If the second derivative becomes large, the denominator increases which slows the convergence (the ratio is the correction to the current position, when the ratio is large, there is little change to the current position and the convergence rate decreases).

A trivial example is the objective function

$$f(x) = x^2 \qquad 2.30$$

The stationary point is at $x = 0$. Suppose we start at the point $x_0 \neq 0$. The iterative function is

$$x_{n+1} = x_n - \frac{2x_n}{2} \qquad 2.31$$

$$x_{n+1} = 0 \qquad 2.32$$

Thus, no matter what point we start at, the next point is always $x = 0$, which is the correct value of the stationary point.

As another example, examine the objective function

$$f(x) = x^4 \qquad 2.33$$

Applying the algorithm we find,

$$x_{n+1} = x_n - \frac{4x_n^3}{12x_n^2} = \frac{2}{3}x_n \qquad 2.34$$

Starting from the point x_0,

$$x_{n+1} = \left(\frac{2}{3}\right)^{n+1} x_0 \qquad 2.35$$

As n gets larger, we converge to the stationary point $x = 0$.

Newton's method may be applied to functions of more than one variable. The principles are essentially unchanged. We assume that we are in the vicinity of a stationary point, approximate the function as a n-dimensional parabola, use the local first and second partial derivatives to determine the form of the parabola, then solve for the stationary point of the parabola. We iterate this process until we achieve sufficient convergence.

One minor difference is that it is common to scale the process by a step size $0 \leq \gamma \leq 1$. Although this can slow the convergence of the algorithm, this helps to keep the process localized to the region of interest. Otherwise, if the second derivatives become small, we can move to an entirely different region of the objective function and find a stationary point far from our original location.

Also, the multivariate version of Newton's method relies on the Hessian matrix. This Hessian acts as a substitute for the second derivative of the function.

Newton's Method I

I-1   Dependent variable z

I-2   Independent variable x

I-3   Single variable, twice differentiable function $f(x)$ relating $z = f(x)$

I-4   Starting point to seed the algorithm x_0

I-5   Uncertainty tolerance ε

S-1   Compute the first derivative of the objective function $f'(x)$

S-3   Compute $x_{n+1} = x_n - \dfrac{f(x_n)}{f'(x_n)}$

S-4   Repeat S-3 until $|x_{n+1} - x_n| < \varepsilon$

O-1 x_{n+1} value of a zero for the objective function

Algorithm IX: Newton's method may be used to find a zero for the objective function.

Newton's Method II

I-1   Dependent variable z

I-2   Independent variables $\vec{x}$

I-3   Multivariable, twice differentiable function $f(\vec{x})$ relating $z = f(\vec{x})$

I-4   Starting point to seed the algorithm $\vec{x}_0$

I-5   Uncertainty tolerance ε

I-6   Step size γ

S-1   Compute the gradient of the objective function $\nabla f(\vec{x})$

S-2   Compute the Hessian of the objective function $Hf(\vec{x})$

S-3   Compute $\vec{x}_{n+1} = \vec{x}_n - \gamma[H^{-1}f(\vec{x}_n)]\nabla f(\vec{x}_n)$

S-4   Repeat S-3 until $|\vec{x}_{n+1} - \vec{x}_n| < \varepsilon$

O-1 $\vec{x}_{n+1}$ value of a stationary point the objective function

Algorithm X: Newton's method may be extended to multivariate objective functions.

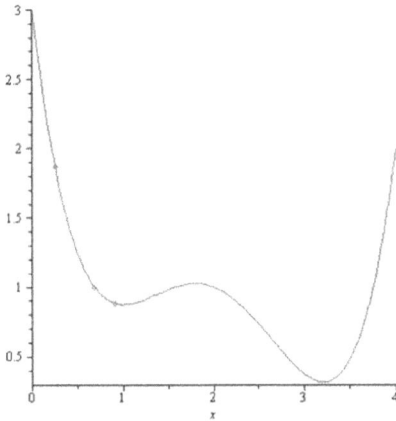

Figure 12: Newton's method converging to the   Figure 13: Newton's method converging to the
stationary point at x=1.   stationary point at (1,1).

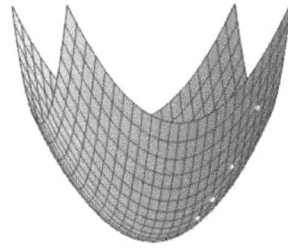

For a 1-dimensional example, examine the objective function

$$f(x) = \frac{x^4}{4} - 2x^3 + \frac{10.75}{2}x^2 - 5.75x + 3 \qquad 2.36$$

The objective function is shown in Figure 12. If we begin at the point $x_0 = .25$, the first few iterations of Newton's method are

$$x_1 = .68 \qquad\qquad 2.37$$

$$x_2 = .91 \qquad\qquad 2.38$$

$$x_3 = .99 \qquad\qquad 2.39$$

converging to the stationary point at $x = 1$.

For a two dimensional example, choose the objective function

$$f(x, y) = (x - 1)^2 + (y - 1)^2 \qquad 2.40$$

This function is shown in Figure 13. The Hessian matrix is

$$Hf(x, y) = \begin{bmatrix} \dfrac{\partial^2 f}{\partial x^2} & \dfrac{\partial^2 f}{\partial x \partial y} \\ \dfrac{\partial^2 f}{\partial y \partial x} & \dfrac{\partial^2 f}{\partial y^2} \end{bmatrix} = 2\begin{bmatrix} 1 & 0 \\ 0 & 1 \end{bmatrix} \qquad 2.41$$

The first few points from Newton's Method are shown as the lighter points in Figure 13. We see that the points converge to the stationary point at $(1,1)$.

Note that the one dimensional Newton's method finds zeros of the function while the multidimensional version is designed to find stationary points.

2.7 Method of Steepest Descents

The method of steepest descents is similar to Newton's method. However, instead of fully approximating the local objective function as a parabola, we simply compute the gradient of the function and move in the direction of the gradient. The idea is if we keep moving downhill, we will eventually reach a stationary point.

As an input, we need a step size γ. The stepsize is allowed to change with each iteration. Variants of the algorithm prescribe different values of γ based on the history of the convergence, magnitude of the gradient, or other factors determined from the objective function.

Similar to Newton's method, the method of steepest descents is an iterative method. However, the method of steepest descents does not use second derivatives, so the Hessian matrix is not computed. The recurrence relation for this method is

$$\vec{x}_{n+1} = \vec{x}_n - \gamma \nabla f(\vec{x}_n) \qquad\qquad 2.42$$

The advantage of this method over Newton's method is that steepest descents does not require the evaluation of the Hessian matrix. This can reduce the computational time required to find the next iteration. However, the disadvantage is that because second derivative information is not present, the number of iterations required for convergence increases.

Steepest Descents

I-1   Dependent variable z

I-2   Independent variables $\vec{x}$

I-3   Multivariable, differentiable function $f(\vec{x})$ relating $z = f(\vec{x})$

I-4   Starting point to seed the algorithm $\vec{x}_0$

I-5   Uncertainty tolerance ε

I-6   Step size γ

S-1   Compute the gradient of the objective function $\nabla f(\vec{x})$

S-2   Compute $\vec{x}_{n+1} = \vec{x}_n - \gamma \nabla f(\vec{x}_n)$

S-3   Repeat S-2 until $|\vec{x}_{n+1} - \vec{x}_n| < \varepsilon$

O-1 $\vec{x}_{n+1}$ value of a stationary point for the objective function

Algorithm XI: Steepest descents may be used to optimize a multivariate objective function.

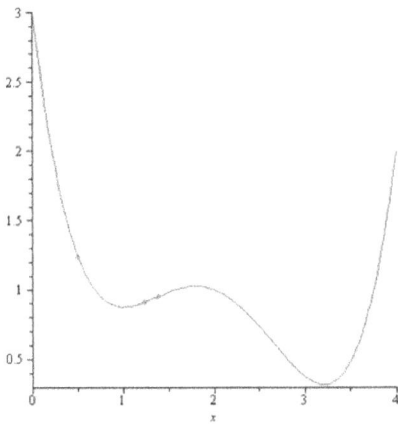

Figure 14: Starting form x=.5, steepest descents converges to the stationary point at x=1.

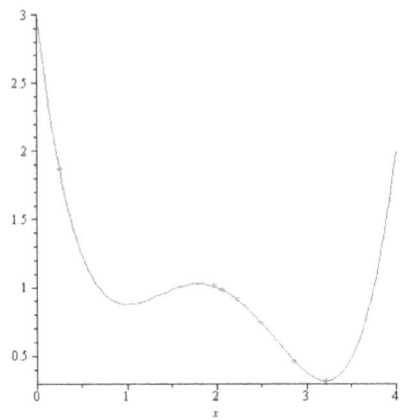

Figure 15: Starting from x=.25, steepest descents converges to the second stationary point rather than the stationary point at x=1.

Let's examine the convergence of steepest descents for the same objective function as we examined under Newton's method. The objective function is

$$f(x) = \frac{x^4}{4} - 2x^3 + \frac{10.75}{2}x^2 - 5.75x + 3 \qquad \text{2.43}$$

In one dimension, the gradient is just the derivative:

$$f'(x) = x^3 - 6x^2 + 10.75x - 5.75 \qquad \text{2.44}$$

Figure 14 shows the convergence of this method to the stationary point $x = 1$ starting from the initial point $x = .5$ with a step size $\gamma = .5$. The first few points in the sequence are

$$x_1 = 1.38 \qquad \text{2.45}$$

$$x_2 = 1.23 \qquad \text{2.46}$$

$$x_3 = 1.10 \qquad \text{2.47}$$

The method overshoots the stationary point, then retreats back. In fact, if we started with the point $x = .25$ (as we did when applying Netwon's method), the method entirely overshoots the stationary point and converges to the other stationary point.

This is a common problem with these methods. Even with an initial value in the vicinity of one stationary point, we can become trapped in another stationary point and converge to the second point. Other methods provide more control, and in fact some methods are able to find both points.

2.8 Bisection Method

If the objective function is differentiable, then zeros of the derivative are the stationary points for the objective function. Examining the derivative, if we know that there is exactly one zero of the derivative between the values x_L and x_R (x-left and x-right), then we may use the bisection method to bracket the zero into smaller and smaller regions.

If a function has exactly one zero on the range $[x_L, x_R]$, then the function must either change signs as it passes through the zero, or the zero must also be an extrema. The bisection method assumes that the function changes signs.

The algorithm proceeds iteratively. First we compute a test point midway between the x-values:

$$x_T = \frac{(x_L + x_R)}{2} \qquad 2.48$$

Next, check the derivative at the test point. If the derivative here has the same sign as x_L, then set $x_L = x_T$. If the derivative has the same sign as x_R, then set $x_R = x_T$. If the derivative is exactly zero, then stop.

Bisection Method

I-1   Dependent variable z

I-2   Independent variables x

I-3   Univariate, function $f(x)$ relating $z = f(x)$

I-4   Starting points that bracket exactly one zero of $f(x)$: x_L and x_R

I-5   Uncertainty tolerance ε

S-1   Compute $x_T = \frac{x_L + x_R}{2}$

S-2   Compute $f(x_T)$

S-3   If $f(x_T)$ and $f(x_L)$ have the same sign, then set $x_L = x_T$

S-4   If $f(x_T)$ and $f(x_R)$ have the same sign, then set $x_R = x_T$

S-5   If $f(x_T) = 0$ then output x_T

S-6   Repeat from S-1 until $|x_L - x_R| < \varepsilon$

O-1   Estimate for the zero of the objective function: $x = \frac{x_L + x_R}{2}$

Algorithm XII: Bisection method is used to find the zero of a function.

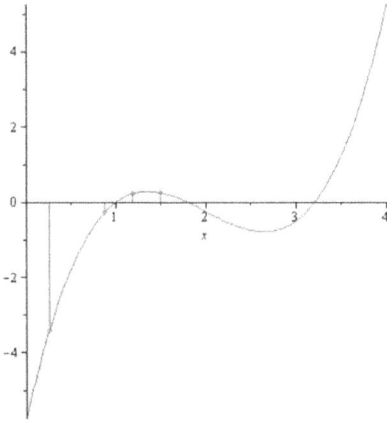

Figure 16: Bisection method converging to the stationary point at x=1. Starting values are x=.25 and x=1.5.

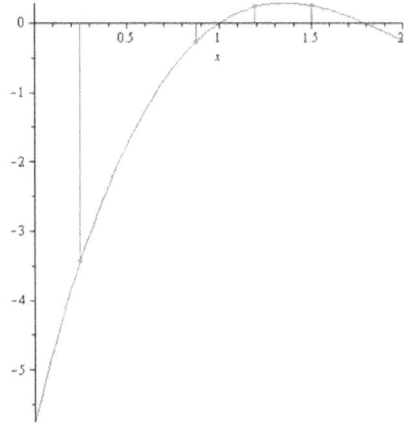

Figure 17: Magnified view of the bisection method from the previous figure.

Examine the objective function

$$f(x) = \frac{x^4}{4} - 2x^3 + \frac{10.75}{2}x^2 - 5.75x + 3 \qquad 2.49$$

The stationary points of the objective function may be found from the zeros of the derivative. The derivative of the objective function is

$$f'(x) = x^3 - 6x^2 + 10.75x - 5.75 \qquad 2.50$$

The points $x = .25$ and $x = 1.5$ are on either side of a zero of the derivative. Applying the bisection method from these starting values we have the sequence

x_L	x_R	$f(x_L)$	$f(x_R)$	x_T	$f(x_T)$	Replace
0.25	1.50	-3.42	0.25	0.88	-0.27	x_L
0.88	1.50	-0.27	0.25	1.19	0.23	x_R
0.88	1.19	-0.27	0.23	1.03	0.05	x_R
0.88	1.03	-0.27	0.05	0.95	-0.09	x_L

The application of this method is shown in Figure 16 and Figure 17. Each iteration of the method reduces the interval for the zero by a factor of two.

The bisection method rapidly converges to the zero of the function. However, we must know two points that bound exactly one zero of the function. When this information is available, the bisection method is useful for finding the zeros of the function.

2.9 Secant Method

The secant method is another method for finding the zeros a function. If our objective function is differentiable, then we may use the secant method to find the zeros of the derivative, which are the stationary points for the objective function.

The secant method is used to find the zeros of a function. We start with two x-values that have different values for the function. We compute the line connecting these points, then find the x-intercept of the line. This x-intercept is used in the next iteration of the method.

The recurrence relation for the secant method is

$$x_{n+1} = x_n - \frac{f(x_n)}{f(x_n) - f(x_{n-1})}(x_n - x_{n-1}) \qquad 2.51$$

The drawback to the secant method is when two consecutive points are near an extremum for the function, then the denominator for the second term can be very small. In this case, x_{n+1} may be very far from x_n spoiling the convergence. The secant method is best applied in regions where the function is not slowly varying.

In another version of the method, we begin with two points on either size of the zero similar to the bisection method. The secant is computed, and the function evaluated at the x-intercept of the secant. In this method, the denominator can never be zero.

Secant Method I

I-1   Dependent variable z

I-2   Independent variables x

I-3   Univariate function $f(x)$ relating $z = f(x)$

I-4   Starting points where the derivative is different: x_0 and x_1, $f(x_0) \neq f(x_1)$

I-5   Uncertainty tolerance ε

S-1   Compute $x_{n+1} = x_n - \frac{f(x_n)}{f(x_n)-f(x_{n-1})}(x_n - x_{n-1})$

S-2   Repeat from S-1 until $|x_n - x_{n-1}| < \varepsilon$

O-1   Estimate for the zero of the objective function: x_{n+1}

Algorithm XIII: Secant method is used to find the zero of a function.

Secant Method II

I-1   Dependent variable z

I-2   Independent variables x

I-3   Univariate function $f(x)$ relating $z = f(x)$

I-4   Starting points where the derivative has opposite signs: x_L and x_R, $f(x_L)f(x_R) < 0$. x_L is to the left of the zero, while x_R is to the right of the zero.

I-5   Uncertainty tolerance ε

S-1   Compute $x_T = x_R - \dfrac{f(x_R)}{f(x_R)-f(x_L)}(x_R - x_L)$

S-2   If $f(x_T)$ and $f(x_L)$ have the same sign, then set $x_L = x_T$

S-3   If $f(x_T)$ and $f(x_R)$ have the same sign, then set $x_R = x_T$

S-4   Repeat from S-1 until $|x_n - x_{n-1}| < \varepsilon$

O-1   Estimate for the zero of the objective function: x_n

Algorithm XIV: Secant method may also be used similar to the bisection method.

By evaluating the function, we can determine if the x-intercept lies to the left or right of the zero. We replace the appropriate point (similar to the bisection method), then iterate. This method also converges to the zero of the function. Figure 18 and Figure 19 show the two secant methods. The first figure illustrates the effect of two consecutive points in a slowly varying region.

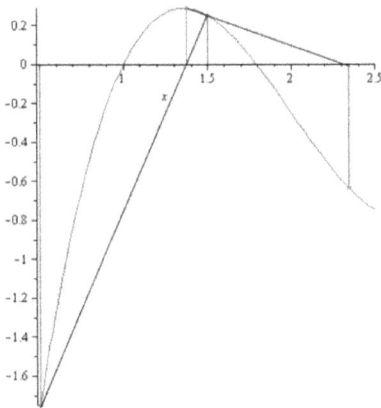

Figure 18: The secant method can move away from the zero if the function is slowly varying between two consecutive points.

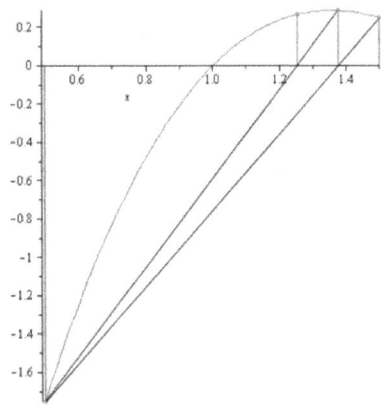

Figure 19: The second secant method always converges to the zero.

2.10   Dekker's Method

Dekker's method is an improvement on the Secant II method by combining the secant method with the bisection method. Dekker's method is a root finding algorithm that identifies a zero of a function given two values of x that lie on either side of the zero.

Let $f(x)$ be a univariate function with exactly one zero in the range (x_L, x_R). Set

$$b_n = \begin{cases} x_L & |f(x_L)| < |f(x_R)| \\ x_R & |f(x_R)| < |f(x_L)| \end{cases} \qquad 2.52$$

and set

$$a_n = \begin{cases} x_L & |f(x_L)| > |f(x_R)| \\ x_R & |f(x_R)| > |f(x_L)| \end{cases} \qquad 2.53$$

The variable b is equal to the value of x where the objective function has the smaller magnitude, while a is the other value. In this sense, b is a better approximation to the true zero than a.

Use the secant to find the x-intercept of the line joining the value of the function at the points b and a:

$$S_T = b_n - \frac{f(b_n)}{f(b_n) - f(b_{n-1})} (b_n - b_{n-1}) \qquad 2.54$$

and the bisection method for the midpoint:

$$B_T = \frac{b_n + a_n}{2} \qquad 2.55$$

One of these points is used in the next iteration. Choose

$$N_T = \begin{cases} S_T & min(b_n, B_T) < S_T < max(b_n, B_T) \\ B_T & otherwise \end{cases} \qquad 2.56$$

Finally, update the appropriate endpoint of the range:

$$x_L = \begin{cases} x_L & f(x_L)f(N_T) < 0 \\ N_T & f(x_L)f(N_T) > 0 \end{cases} \qquad 2.57$$

$$x_R = \begin{cases} x_R & f(x_R)f(N_T) < 0 \\ N_T & f(x_R)f(N_T) > 0 \end{cases} \qquad 2.58$$

These last two statements simply tells us to set either x_L or x_R to N_T. We set $x_L = N_T$ when $f(x_L)$ and $f(N_T)$ have the same sign, and we set $x_R = N_T$ when $f(x_R)$ and $f(N_T)$ have the same sign. Then we repeat the process again, choosing a and b according to 2.52 and 2.53.

Dekker's Method

I-1   Dependent variable z

I-2   Independent variables x

I-3   Univariate function $f(x)$ relating $z = f(x)$

I-4   Starting points where the derivative has opposite signs: x_L and x_R, $f(x_L)f(x_R) < 0$. x_L is to the left of the zero, while x_R is to the right of the zero.

I-5   Uncertainty tolerance ε

S-1   Set $b_n = \begin{cases} x_L & |f(x_L)| < |f(x_R)| \\ x_R & |f(x_R)| < |f(x_L)| \end{cases}$

S-2   Set $a_n = \begin{cases} x_L & |f(x_L)| > |f(x_R)| \\ x_R & |f(x_R)| > |f(x_L)| \end{cases}$

S-3   Compute $S_T = b_n - \dfrac{f(b_n)}{f(b_n) - f(b_{n-1})}(b_n - b_{n-1})$

S-4   Compute $B_T = \dfrac{b_n + a_n}{2}$

S-5   Set $N_T = \begin{cases} S_T & min(b_n, B_T) < S_T < max(b_n, B_T) \\ B_T & otherwise \end{cases}$

S-6   If $f(N_T)$ and $f(x_L)$ have the same sign, then set $x_L = N_T$

S-3   If $f(N_T)$ and $f(x_R)$ have the same sign, then set $x_R = N_T$

S-4   Repeat from S-1 until $|b_n - b_{n-1}| < \varepsilon$

O-1   Estimate for the zero of the objective function: b_n

Algorithm XV: Dekker's method combines the bisection and secant methods.

Dekker's method combines the secant and bisection methods. This method is generally robust for most well behaved functions. However, there are situations where Dekker's method chooses the secant method (S_T) at every iteration and actually converges more slowly than the bisection method.

Brent's method modifies Dekker's method to identify and resolve these issues. Brent's method adds additional criteria for choosing between S_T and B_T at each iteration by examining which method was chosen in the previous iteration.

By examining the past iteration, Brent's method is able to prevent the algorithm from constantly choosing the secant method when the bisection method converges faster.

2.11 Brent's Method

Brent's method is an improvement on Dekker's method. This method computes the zero of a function given two values of x that lie on either side of the zero.

Brent's method is substantially similar to Dekker's method but modifies how the method chooses between using the secant and bisection methods. Brent adds additional criteria that examines previous iterations to assure that the secant method is not constantly chosen. In some situations, Dekker's method consistently chooses the secant method when in fact the bisection method converges faster. Brent's criteria attempt to avoid this difficulty.

Let $f(x)$ be a univariate function with exactly one zero in the range (x_L, x_R). Set

$$b_n = \begin{cases} x_L & |f(x_L)| < |f(x_R)| \\ x_R & |f(x_R)| < |f(x_L)| \end{cases} \qquad 2.59$$

and set

$$a_n = \begin{cases} x_L & |f(x_L)| > |f(x_R)| \\ x_R & |f(x_R)| > |f(x_L)| \end{cases} \qquad 2.60$$

The variable b is equal to the value of x where the objective function has the smaller magnitude, while a is the other value. In this sense, b is a better approximation to the true zero than a.

Use the secant to find the x-intercept of the line joining the value of the function at the points b and a:

$$S_T = b_n - \frac{f(b_n)}{f(b_n) - f(b_{n-1})} (b_n - b_{n-1}) \qquad 2.61$$

and the bisection method for the midpoint:

$$B_T = \frac{b_n + a_n}{2} \qquad 2.62$$

One of these points is used in the next iteration. Choose

$$N_T = \begin{cases} S_T & min(b_n, B_T) < S_T < max(b_n, B_T) \\ B_T & otherwise \end{cases} \qquad 2.63$$

Up to this point, Brent's method is the same as Dekker's method. If $N_T = S_T$ from 2.63 (Dekker's method chose the secant), two inequalities are checked. If the previous iteration used the bisection method, then the current iteration may use the secant method if

$$|\delta| < |b_n - b_{n-1}|$$ 2.64

$$|S_T - b_n| < \frac{|b_n - b_{n-1}|}{2}$$ 2.65

where δ is a numerical tolerance. Both inequalities must hold. If either of these fails, then we set $N_T = B_T$ instead of $N_T = S_T$.

If the previous iteration used the secant method, then the inequalities

$$|\delta| < |b_{n-1} - b_{n-2}|$$ 2.66

$$|S_T - b_n| < \frac{|b_{n-1} - b_{n-2}|}{2}$$ 2.67

must hold. Again, if either of these inequalities fails, we set $N_T = B_T$ for this iteration.

With this correction, we continue as before with Dekker's method. The final step is updating the endpoint of the range according to N_T:

$$x_L = \begin{cases} x_L & f(x_L)f(N_T) < 0 \\ N_T & f(x_L)f(N_T) > 0 \end{cases}$$ 2.68

$$x_R = \begin{cases} x_R & f(x_R)f(N_T) < 0 \\ N_T & f(x_R)f(N_T) > 0 \end{cases}$$ 2.69

Essentially, we either set x_L or x_R to N_T depending on the sign of $f(N_T)$. When $f(x_L)$ and $f(N_T)$ have the same sign we set $x_L = N_T$, and when $f(x_R)$ and $f(N_T)$ have the same sign, we set $x_R = N_T$. The process is repeated iteratively.

As an example, use the objective function

$$f(x) = \frac{x^4}{4} - 2x^3 + \frac{10.75}{2}x^2 - 5.75x + 3$$ 2.70

with the starting points $x_L = .25$ and $x_R = 1.5$. The first few points are:

x_L	x_R	$f(x_L)$	$f(x_R)$	a	b	$f(a)$	$f(b)$	S_T	$f(S_T)$	B_T	$f(B_T)$
0.25	1.50	-3.42	0.25	0.25	0.25	-3.42	-3.42	0.00	-5.75	0.88	-0.27
0.25	1.50	-3.42	0.25	0.25	1.50	-3.42	0.25	1.41	0.28	0.88	-0.27
0.88	1.50	-0.27	0.25	0.88	1.50	-0.27	0.25	1.20	0.24	1.19	0.23
0.88	1.19	-0.27	0.23	0.88	1.19	-0.27	0.23	1.04	0.07	1.03	0.05
0.88	1.03	-0.27	0.05	0.88	1.03	-0.27	0.05	1.01	0.01	0.95	-0.09
0.95	1.03	-0.09	0.05	0.95	1.03	-0.09	0.05	1.00	0.00	0.99	-0.01
0.99	1.03	-0.01	0.05	1.03	0.99	0.05	-0.01	1.00	0.00	1.01	0.02
0.99	1.01	-0.01	0.02	1.01	0.99	0.02	-0.01	1.00	0.00	1.00	0.00

Brent's Method

I-1   Dependent variable z

I-2   Independent variables x

I-3   Univariate function $f(x)$ relating $z = f(x)$

I-4   Starting points where the derivative has opposite signs: x_L and x_R, $f(x_L)f(x_R) < 0$. x_L is to the left of the zero, while x_R is to the right of the zero.

I-5   Uncertainty tolerance ε

S-1   Set $b_n = \begin{cases} x_L & |f(x_L)| < |f(x_R)| \\ x_R & |f(x_R)| < |f(x_L)| \end{cases}$

S-2   Set $a_n = \begin{cases} x_L & |f(x_L)| > |f(x_R)| \\ x_R & |f(x_R)| > |f(x_L)| \end{cases}$

S-3   Compute $S_T = b_n - \dfrac{f(b_n)}{f(b_n)-f(b_{n-1})}(b_n - b_{n-1})$

S-4   Compute $B_T = \dfrac{b_n+a_n}{2}$

S-5   Set $N_T = \begin{cases} S_T & min(b_n, B_T) < S_T < max(b_n, B_T) \\ B_T & otherwise \end{cases}$

S-6   If $N_T = S_T$, and if the previous iteration used B_T, and either
$$|\delta| < |b_n - b_{n-1}|$$
$$|S_T - b_n| < \frac{|b_n - b_{n-1}|}{2}$$
are not satisfied, then set $N_T = S_T$.

S-7   If $N_T = S_T$, and if the previous iteration used S_T, and either
$$|\delta| < |b_{n-1} - b_{n-2}|$$
$$|S_T - b_{n-1}| < \frac{|b_{n-1} - b_{n-2}|}{2}$$
are not satisfied, then set $N_T = S_T$.

S-6   If $f(N_T)$ and $f(x_L)$ have the same sign, then set $x_L = N_T$

S-3   If $f(N_T)$ and $f(x_R)$ have the same sign, then set $x_R = N_T$

S-4   Repeat from S-1 until $|b_n - b_{n-1}| < \varepsilon$

O-1   Estimate for the zero of the objective function: b_n

Algorithm XVI: Dekker's method combines the bisection and secant methods.

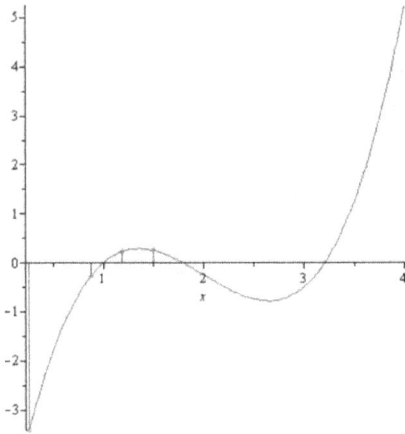

Figure 20: Brent's method is a combination of the secant and bisection methods.

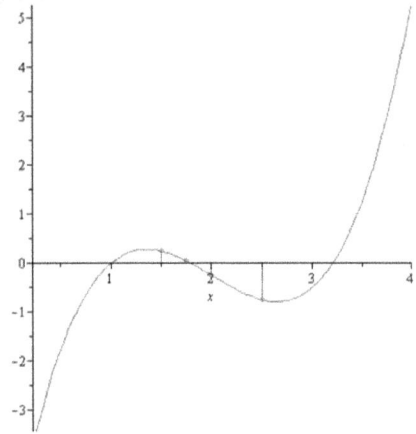

Figure 21: Brent's method converging to the other zero of the objective function.

The convergence of Brent's method to the zero of the objective function is shown in Figure 20. If we choose the initial points $x_L = 1.5$ and $x_R = 2.5$, Brent's method converges to the other zero of the objective function as shown in Figure 21. The first few iterations in this case are:

x_L	x_R	$f(x_L)$	$f(x_R)$	a	b	$f(a)$	$f(b)$	S_T	$f(S_T)$	B_T	$f(B_T)$
1.50	2.50	0.25	-0.75	2.50	0.25	-0.75	-3.42	0.00	-5.75	2.00	-0.25
1.50	2.50	0.25	-0.75	2.50	1.50	-0.75	0.25	1.41	0.28	2.00	-0.25
1.50	2.00	0.25	-0.25	2.00	1.50	-0.25	0.25	1.75	0.05	1.75	0.05
1.75	2.00	0.05	-0.25	2.00	1.75	-0.25	0.05	1.79	0.00	1.88	-0.10
1.75	1.88	0.05	-0.10	1.88	1.75	-0.10	0.05	1.79	0.00	1.81	-0.02
1.75	1.81	0.05	-0.02	1.75	1.81	0.05	-0.02	1.79	0.00	1.78	0.01
1.78	1.81	0.01	-0.02	1.81	1.78	-0.02	0.01	1.79	0.00	1.80	0.00
1.78	1.80	0.01	0.00	1.78	1.80	0.01	0.00	1.79	0.00	1.79	0.00

This case is interesting because at the third iteration

$$|f(x_L)| = |f(x_R)| \qquad \text{2.71}$$

At this point, the process no longer meets the requirement for choosing a and b in steps S-1 and S-2.

When $|f(x_L)| = |f(x_R)|$, the denominator of the secant term in step S-3 becomes zero. In this case, we simply use the bisection method for this iteration, then return to the normal Brent's method for subsequent iterations.

Although it is rare that $|f(x_L)| = |f(x_R)|$ is exactly met, we should be aware of the possibility for this and design the algorithm accordingly. The only modification needed is to use the bisection method in this case.

2.12   Halley's Method

Halley's method is another method for computing a zero of a function. The method is similar to Newton's method. In fact, the two methods belong to the same class of methods called Householder's methods.

Similar to Newton's method, Halley's method requires an initial starting point x_0. If we have prior knowledge of where a root might be located, choosing x_0 near this point is helpful. However, it is not required that x_0 be in the vicinity of the root.

The algorithm is computed iteratively as

$$x_{n+1} = x_n - \frac{2f(x_n)f'(x_n)}{2[f'(x_n)]^2 - f(x_n)f''(x_n)}$$
(2.72)

The connection with Newton's method is more apparent if we write this in the form

$$x_{n+1} = x_n - \frac{f'(x_n)}{f''(x_n)}\left[1 - \frac{f(x_n)f''(x_n)}{2[f'(x_n)]^2}\right]^{-1}$$
(2.73)

From this perspective, Halley's method multiplies the root finding term by a correction factor. Generally, Haley's method converges cubically to the root. This outperforms Newton's method which converges quadratically.

Halley's Method

I-1	Dependent variable z
I-2	Independent variable x
I-3	Single variable, twice differentiable function $f(x)$ relating $z = f(x)$
I-4	Starting point to seed the algorithm x_0
I-5	Uncertainty tolerance ε
S-1	Compute the first derivative of the objective function $f'(x)$
S-2	Compute the second derivative of the objective function $f''(x)$
S-3	Compute $x_{n+1} = x_n - \frac{2f(x_n)f'(x_n)}{2[f'(x_n)]^2 - f(x_n)f''(x_n)}$
S-4	Repeat S-3 until $|x_{n+1} - x_n| < \varepsilon$
O-1	x_{n+1} value of a zero for the objective function

Algorithm XVII: Halley's method may be used to find a zero for the objective function.

Halley's method may be derived from Newton's method by applying Newton's method to the function

$$g(x) = \frac{f(x)}{\sqrt{|f'(x)|}} \qquad\qquad 2.74$$

So long as the function does not have zeros that coincide with the derivative, each zero of the $f(x)$ is also a zero of $g(x)$. Moreover, if the derivative is everywhere finite, each zero of $g(x)$ is also a zero of $f(x)$.

Applying Newton's method to $g(x)$,

$$x_{n+1} = x_n - \frac{g(x_n)}{g'(x_n)} \qquad\qquad 2.75$$

The derivatives of $g(x)$ may be written in terms of $f(x)$ and its derivatives:

$$g'(x) = \frac{f'(x)\sqrt{|f'(x)|} - \frac{f(x)f''(x)}{2\sqrt{|f'(x)|}}sgn(f'(x))}{|f'(x)|} \qquad\qquad 2.76$$

$$= \frac{2f'(x) - \frac{f(x)f''(x)sgn(f'(x))}{|f'(x)|}}{2\sqrt{|f'(x)|}} \qquad\qquad 2.77$$

$$= \frac{2f'(x)|f'(x)| - f(x)f''(x)sgn(f'(x))}{2|f'(x)|\sqrt{|f'(x)|}} \qquad\qquad 2.78$$

$$= \frac{2[f'(x)]^2 sgn(f'(x)) - f(x)f''(x)sgn(f'(x))}{2f'(x)sgn(f'(x))\sqrt{|f'(x)|}} \qquad\qquad 2.79$$

$$= \frac{2[f'(x)]^2 - f(x)f''(x)}{2f'(x)\sqrt{|f'(x)|}} \qquad\qquad 2.80$$

Thus,

$$\frac{g(x_n)}{g'(x_n)} = \frac{f(x)}{\sqrt{|f'(x)|}}\frac{2f'(x)\sqrt{|f'(x)|}}{2[f'(x)]^2 - f(x)f''(x)} \qquad\qquad 2.81$$

$$= \frac{2f'(x)f(x)}{2[f'(x)]^2 - f(x)f''(x)} \qquad\qquad 2.82$$

Substituting into the iterative equation for Newton's method,

$$x_{n+1} = x_n - \frac{2f'(x_n)f(x_n)}{2[f'(x_n)]^2 - f(x_n)f''(x_n)} \qquad\qquad 2.83$$

2.13   Householder's Methods

Householder's methods are a set of methods that generalize Newton's method and Haley's method. Whereas Newton's method requires the first derivative and Halley's required the second derivative, the kth order Householder method required the kth order derivative. However, as Newton's method converges as n^2 and Halley's method converges as n^3, Householders method converges as n^k. By increasing the number of derivatives involved in the iterative sequence, we can increase the rate of convergence to a zero.

The method begins with an initial point x_0. The nth order approximation to the zero is given by

$$x_{n+1} = x_n + k \frac{\frac{d^{k-1}}{dx^{k-1}}(1/f(x_n))}{\frac{d^k}{dx^k}(1/f(x_n))} \tag{2.84}$$

Generally, the computation of k derivatives of $1/f(x)$ is very tedious. However, the first few methods can be computed directly. For example, the first order Householder method is

$$x_{n+1} = x_n + \frac{(1/f(x_n))}{\frac{d}{dx}(1/f(x_n))} \tag{2.85}$$

$$= x_n + \frac{(1/f(x_n))}{-(f'(x_n)/f^2(x_n))} \tag{2.86}$$

$$= x_n - \frac{f(x_n)}{f'(x_n)} \tag{2.87}$$

This is the same as Newton's method. For the Householder method at second order, we have

$$\frac{d}{dx}(1/f(x_n)) = -\frac{f'(x)}{f^2(x)} \tag{2.88}$$

and

$$\frac{d^2}{dx^2}(1/f(x_n)) = \frac{2[f'(x)]^2 - f(x)f''(x)}{[f(x)]^3} \tag{2.89}$$

Thus,

$$x_{n+1} = x_n - 2\frac{f'(x)}{f^2(x)}\frac{[f(x)]^3}{2[f'(x)]^2 - f(x)f''(x)} \tag{2.90}$$

Householder's Method

I-1   Dependent variable z

I-2   Independent variable x

I-3   Single variable, k times differentiable function $f(x)$ relating $z = f(x)$

I-4   Starting point to seed the algorithm x_0

I-5   Uncertainty tolerance ε

S-1   Compute the first k derivatives of $1/f(x)$

S-3   Compute $x_{n+1} = x_n + k \dfrac{\frac{d^{k-1}}{dx^{k-1}}(1/f(x_n))}{\frac{d^k}{dx^k}(1/f(x_n))}$

S-4   Repeat S-3 until $|x_{n+1} - x_n| < \varepsilon$

O-1 x_{n+1} value of a zero for the objective function

Algorithm XVIII: Householder's method may be used to find a zero for the objective function.

$$x_{n+1} = x_n - \frac{2f'(x)f(x)}{f(x)f''(x) - 2[f'(x)]^2} \qquad 2.91$$

This is the iterative equation for Halley's method. The process may be continued for higher orders. The table below shows the first few derivatives of $1/f(x)$. The next table shows the first five Householder methods. Computing the Householder methods becomes increasingly difficult at higher orders.

Order	$\dfrac{d^k}{dx^k}\left(1/f(x)\right)$
0	$1/f(x)$
1	$-f'/f^2$
2	$-(f'' - 2f'^2)/f^3$
3	$-(6f'^3 - 6ff'f'' + f^2f''')/f^4$
4	$-\left(-24f'^4 + 36ff'^2f'' - 6f^2f''^2 - 8f^2f'f''' + f^3f^{(4)}\right)/f^5$
5	$-\left(120f'^5 - 240ff'^3f'' + 90f^2f'f''^2 + 60f^2f'^2f''' \\ - 20f^3f''f''' - 10f^3f'f^{(4)} + f^4f^{(5)}\right)/f^6$

Order	Householder's Method
1	$$x_{n+1} = x_n - \frac{f}{f'}$$
2	$$x_{n+1} = x_n - \frac{2ff'}{2f'^2 - ff''}$$
3	$$x_{n+1} = x_n - \frac{f(6f'^2 - 3ff'')}{6f'^3 - 6ff'f'' + f^2f'''}$$
4	$$x_{n+1} = x_n - \frac{f(24ff'f'' - 24f'^3 - 4f^2f''')}{36ff'^2f'' - 24f'^4 - 6f^2f''^2 - 8f^2f'f''' + f^3f^{(4)}}$$
5	$$x_{n+1} = x_n - \frac{f(120f'^4 - 180ff'^2f'' + 30f^2f''^2 + 40f^2f'f''' - 5f^3f^{(4)})}{120f'^5 - 240ff'^3f'' + 90f^2f'f''^2 + 60f^2f'^2f''' - 20f^3f''f''' - 10f^3f'f^{(4)} + f^4f^{(5)}}$$

As an example, examine the objective function

$$f(x) = (x - .5)(x - 1)(x - 1.5)(x - 2)(x - 2.25)(x - 3) \qquad 2.92$$

A graph of this function is shown in Figure 24. The table below provides the first few iterations of five Householder methods. The convergence of these are shown in Figure 25-Figure 29.

	1	2	3	4	5
0	0.750000000	0.750000000	0.750000000	0.750000000	0.750000000
1	1.058219150	0.892343358	1.108068805	0.916197016	1.116374762
2	0.988232501	0.993636904	0.994115851	0.994115851	0.999151386
3	0.999705178	0.999998110	0.999999985	1.000000001	1.000000000

Higher order Householder methods converge rapidly to the zero of the function.

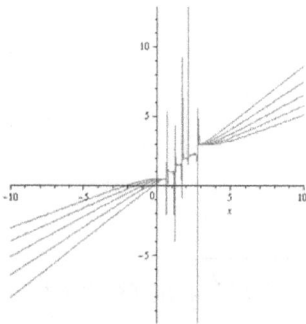

Figure 22: Five iterations of the 1st Householder method.

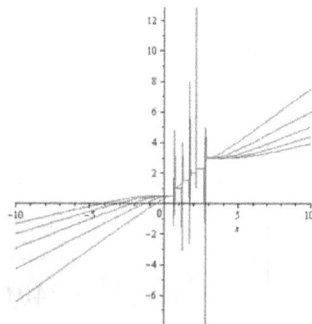

Figure 23: Convergence of the second iteration of the first five Householder methods.

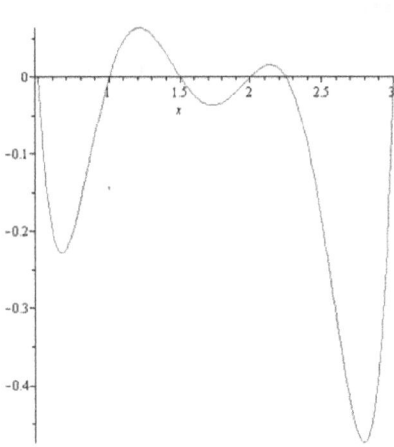

Figure 24: Objective function.

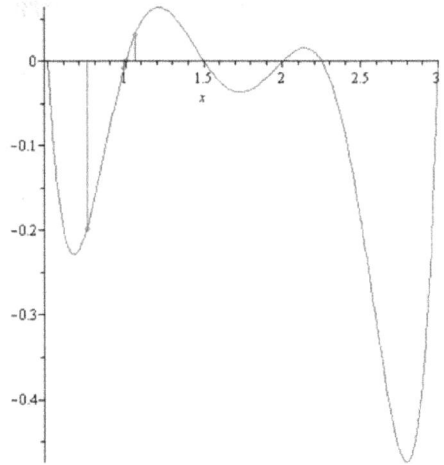

Figure 25: Householder's 1st method.

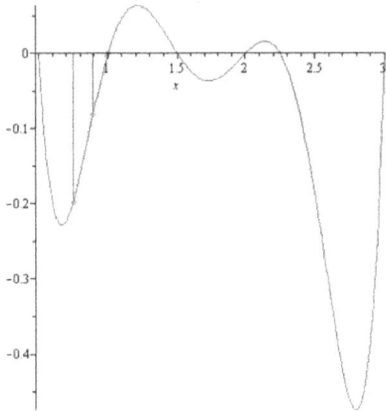

Figure 26: Householder's 2nd method.

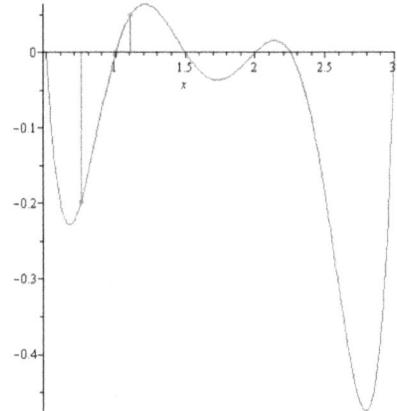

Figure 27: Householder's 3rd method.

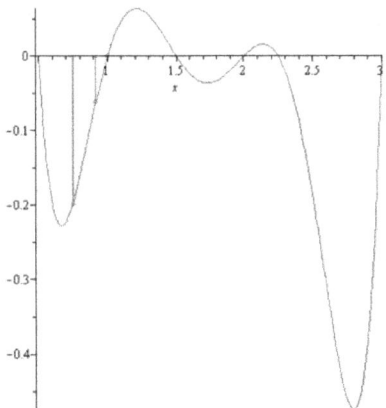

Figure 28: Householder's 4th method.

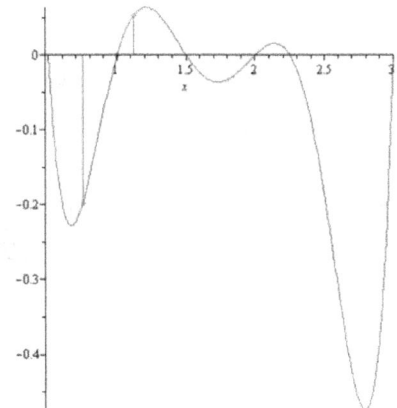

Figure 29: Householder's 5th method.

2.14   Inverse Quadratic Interpolation

Inverse quadratic interpolation is an iterative root-finding method. This method begins with three initial x-values x_0, x_1, and x_2. The value of the objective function is computed at each of these points. Set

$$f_0 = f(x_0) \qquad\qquad 2.93$$

$$f_1 = f(x_1) \qquad\qquad 2.94$$

$$f_2 = f(x_2) \qquad\qquad 2.95$$

Generally, set

$$f_n = f(x_n) \qquad\qquad 2.96$$

From these three value of the function, we compute the approximation to the generalized quadratic. The fit is solved for one of the roots of the quadratic, and this point is used at the next value in the sequence.

The recurrence relation is

$$x_{n+1} = \frac{f_{n-1}f_n}{(f_{n-2}-f_{n-1})(f_{n-2}-f_n)}x_{n-2} + \frac{f_{n-2}f_n}{(f_{n-1}-f_{n-2})(f_{n-1}-f_n)}x_{n-1} + \frac{f_{n-2}f_{n-1}}{(f_n-f_{n-2})(f_n-f_{n-1})}x_n \qquad 2.97$$

Inverse Quadratic Interpolation

I-1   Dependent variable z

I-2   Independent variable x

I-3   Single variable function $f(x)$ relating $z = f(x)$

I-4   Three starting points to seed the algorithm x_0, x_1, and x_2.

I-5   Uncertainty tolerance ε

Compute

S-1 $$x_{n+1} = \frac{f_{n-1}f_n}{(f_{n-2}-f_{n-1})(f_{n-2}-f_n)}x_{n-2} + \frac{f_{n-2}f_n}{(f_{n-1}-f_{n-2})(f_{n-1}-f_n)}x_{n-1} + \frac{f_{n-2}f_{n-1}}{(f_n-f_{n-2})(f_n-f_{n-1})}x_n$$

S-2   Repeat S-1 until $|x_{n+1} - x_n| < \varepsilon$

O-1 x_{n+1} value of a zero for the objective function

Algorithm XIX: Inverse quadratic interpolation is used to find a zero for the objective function.

As an example, examine the objective function

$$f(x) = (x - .5)(x - 1)(x - 1.5)(x - 2)(x - 2.25)(x - 3) \qquad 2.98$$

A graph of this function is shown in Figure 30. The table below shows the sequence starting from the points $x_0 = 0.7$, $x_1 = 0.9$, and $x_2 = 1.3$. The three initial points are all in the vicinity of the zero at $x = 1$. From the table below (Point 1), we see that the algorithm does converge to the zero at $x = 1$. However, the points in the sequence begin by leaving the vicinity of the root, then eventually settle down and converge to $x = 1$.

Alternatively, if we begin with the points $x_0 = 1.8$, $x_0 = 1.9$, and $x_2 = 2.1$ (Point 2), the convergence is much faster. In this case the algorithm quickly converges to the zero at $x = 2$.

Iteration	Point 1	Point 2
0	0.700000000	1.800000000
1	0.900000000	1.900000000
2	1.300000000	2.100000000
3	1.106294728	2.021381026
4	0.001515680	1.995901637
5	-0.750274270	1.999719531
6	1.112061203	2.000000374
7	1.118053118	2.000000000
8	0.929143300	2.000000000

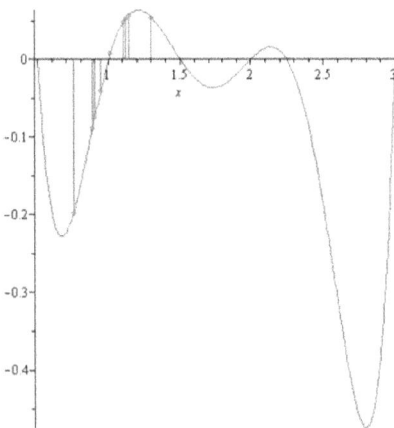

Figure 30: Inverse quadratic interpolation converging to the zero at $x = 1$.

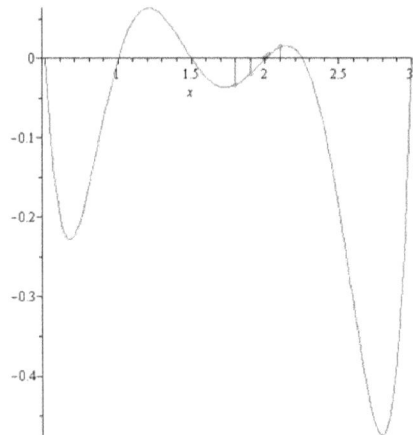

Figure 31: Inverse quadratic interpolation converging to the zero at $x = 2$.

2.15   Müller's Method

Müller's method is another parabolic interpolation method similar to inverse quadratic interpolation. Again, we start with an objective function $f(x)$ and three initial x values:

$$f_0 = f(x_0) \qquad \text{2.99}$$

$$f_1 = f(x_1) \qquad \text{2.100}$$

$$f_2 = f(x_2) \qquad \text{2.101}$$

More generally,

$$f_n = f(x_n) \qquad \text{2.102}$$

The recurrence relation for Müller's method is

$$x_{n+1} = x_n - \frac{2f(x_n)}{w \pm \sqrt{w^2 - 4f(x_n)h}} \qquad \text{2.103}$$

where

$$w = \frac{f_n - f_{n-1}}{x_n - x_{n-1}} + \frac{f_n - f_{n-2}}{x_n - x_{n-2}} - \frac{f_{n-1} - f_{n-2}}{x_{n-1} - x_{n-2}} \qquad \text{2.104}$$

$$h = \frac{\dfrac{f_n - f_{n-1}}{x_n - x_{n-1}} - \dfrac{f_{n-1} - f_{n-2}}{x_{n-1} - x_{n-2}}}{x_n - x_{n-2}} \qquad \text{2.105}$$

Examine the objective function

$$f(x) = (x - .5)(x - 1)(x - 1.5)(x - 2)(x - 2.25)(x - 3) \qquad \text{2.106}$$

A graph of this function is shown in Figure 32 and Figure 33, along with the convergence of the method to the points $x = 1$ and $x = 2$.

Iteration	Point 1	Point 2
0	0.700000000	1.800000000
1	0.900000000	1.900000000
2	1.300000000	2.100000000
3	1.051208941	1.997697361
4	0.999036409	1.999951413
5	1.000001513	1.999719531
6	0.999999999	2.000000023
7	1.000000000	2.000000000
8	1.000000000	2.000000000

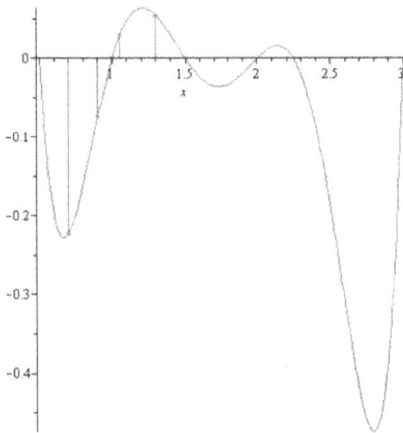

Figure 32: Müller's method converging to the zero at $x = 1$.

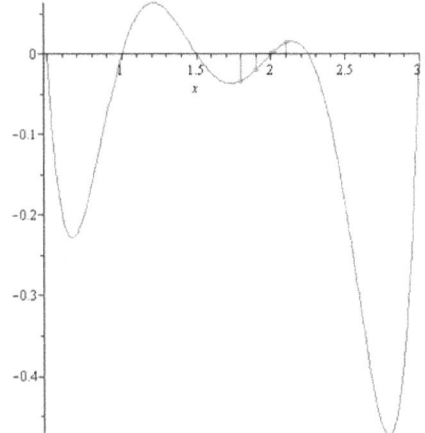

Figure 33: Müller's method converging to the zero at $x = 2$.

From the graphs and the sequence values in the tables, the method quickly converges to the zeros of the function. Generally, this method converges faster than the secant method but not quite as fast as Newton's method.

Müller's Method

I-1   Dependent variable z

I-2   Independent variable x

I-3   Single variable function $f(x)$ relating $z = f(x)$

I-4   Three starting points to seed the algorithm x_0, x_1, and x_2.

I-5   Uncertainty tolerance ε

S-1   Compute $w = \dfrac{f_n - f_{n-1}}{x_n - x_{n-1}} + \dfrac{f_n - f_{n-2}}{x_n - x_{n-2}} - \dfrac{f_{n-1} - f_{n-2}}{x_{n-1} - x_{n-2}}$

S-2   Compute $h = \dfrac{\dfrac{f_n - f_{n-1}}{x_n - x_{n-1}} - \dfrac{f_{n-1} - f_{n-2}}{x_{n-1} - x_{n-2}}}{x_n - x_{n-2}}$

S-3   Compute $x_{n+1} = x_n - \dfrac{2f(x_n)}{w \pm \sqrt{w^2 - 4f(x_n)h}}$

S-4   Repeat from S-1 until $|x_{n+1} - x_n| < \varepsilon$

O-1 x_{n+1} value of a zero for the objective function

Algorithm XX: Müller's method is used to find a zero for the objective function.

2.16   Steffensen's Method

Steffensen's method is a univariate root finding method that does not require computing the derivative of the objective function. Instead, the method requires that in the vicinity of the root the objective function obeys

$$-1 < f'(x) < 0 \qquad \text{2.107}$$

In the vicinity of the root, the objective function must be negatively sloped, and not too steep. Steffessen's method may converge when this criterion is not met, but this is required for optimal performance.

The recurrence relation is

$$x_{n+1} = x_n - \frac{f^2(x_n)}{f\big(x_n + f(x_n)\big) - f(x_n)} \qquad \text{2.108}$$

When the condition 2.107 is met, Steffensen's method converges quadratically similar to Newton's method. The main advantage of Steffensen's method over Newton's method is that Steffensen's method does not require computing the derivative of the objective function.

The largest drawback is adherence to the criterion from 2.107. In many cases, the objective function may not meet this criterion, and even verifying that the condition is met requires computation of the derivative of the objective function. This partially defeats the main advantage that Steffensen's method does not require computing the derivative of the objective function. For these reasons, Steffensen's method is less preferred to other root finding algorithms.

Steffensen's Method

I-1	Dependent variable z
I-2	Independent variable x
I-3	Single variable function $f(x)$ relating $z = f(x)$
I-4	Starting points to seed the algorithm x_0.
I-5	Uncertainty tolerance ε
S-1	Compute $x_{n+1} = x_n - \frac{f^2(x_n)}{f(x_n + f(x_n)) - f(x_n)}$
S-3	Repeat from S-1 until $\lvert x_{n+1} - x_n \rvert < \varepsilon$
O-1	x_{n+1} value of a zero for the objective function

Algorithm XXI: Steffessen's method is used to find a zero for the objective function.

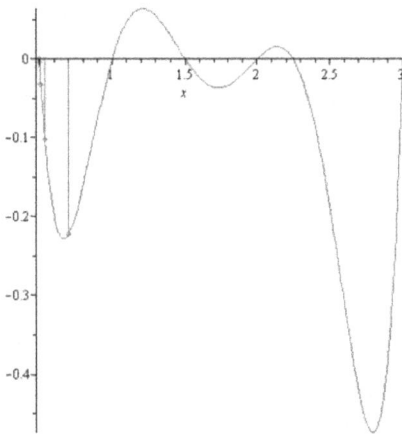

Figure 34: Steffensen's method converging to the zero at $x = .5$.

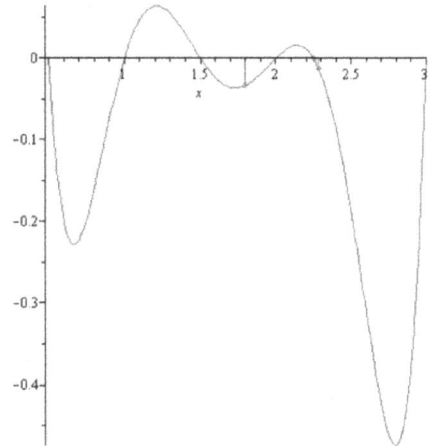

Figure 35: Müller's method converging to the zero $x = 2.25$.

Using the same example objective function from the previous sections,

$$f(x) = (x - .5)(x - 1)(x - 1.5)(x - 2)(x - 2.25)(x - 3) \qquad 2.109$$

We apply Steffensen's method starting from the point $x = .7$. From this initial point, repeated application of Steffensen's method converges to the point $x = .5$. From Figure 34, we see that the objective function has negative slope near the point $x = .5$. However, the slope here is > -3 so 2.107 is not met. Steffensen's method may converge when this condition is not met, but performance is not guaranteed.

We also apply the method starting from the point $x = 1.8$. The convergence in this case is shown in Figure 35. The method converges to the zero at $x = 2.25$. Although the starting point is closer to the zero at $x = 2$, the method converges to the negatively sloped zero at $x = 2.25$.

This is not a general feature. When the initial point is closer to the zero at $x = 2$, the method does in fact converge to $x = 2$. For example, the initial points $x = 1.9$ and x=1.95 both converge to the zero at $x = 2$.

Iteration	Point 1	Point 2
0	0.700000000	1.800000000
1	0.537273082	2.285014049
2	0.510250165	2.253294525
3	0.501001064	2.250039049
4	0.500010493	2.250000007

2.17   Relaxation

Relaxation is a method for optimizing multivariate objective functions by examining each variable in isolation. We begin with a multivariate objective function $f(\vec{x})$ and compute the partial derivatives $\frac{\partial f(\vec{x})}{\partial x_k}$. The stationary points are located at the values of $\vec{x}$ where the partial derivatives are all zero

$$\frac{\partial f(\vec{x})}{\partial x_k} = 0 \qquad\qquad 2.110$$

Relaxation iteratively finds such a point. We begin at an initial point $\vec{x}^0$. For each iteration we minimize the objective function varying each component separately. Starting from the point $\vec{x}^0 = (x_1^0, x_2^0, ..., x_n^0)$, find a component x_1^1 such that the point $(x_1^1, x_2^0, ..., x_n^0)$ minimizes the partial derivative with respect to x_1:

$$\frac{\partial f\left(\vec{x} = (x_1^1, x_2^0, ..., x_n^0)\right)}{\partial x_1} = 0 \qquad\qquad 2.111$$

Next, find a component x_2^1 such that the point $(x_1^1, x_2^1, ..., x_n^0)$ minimizes the partial derivative with respect to x_2:

$$\frac{\partial f\left(\vec{x} = (x_1^1, x_2^1, ..., x_n^0)\right)}{\partial x_2} = 0 \qquad\qquad 2.112$$

This process is continued for each of the components until we have a new vector

$$\vec{x}^1 = (x_1^1, x_2^1, ..., x_n^1) \qquad\qquad 2.113$$

This is the point we use to seed the next iteration.

A common modification to this method is to minimize the components in a random order at each iteration. Thus, at the beginning of the iteration we choose a random order for minimizing the components. This helps to prevent the algorithm from getting stuck in a repetitive cycle.

The major drawback with the relaxation method is that it is not guaranteed to converge to a stationary point. The algorithm may become trapped in cycles of repeating values of $\vec{x}$. By choosing the order for minimizing randomly, this may be avoided. However, there is still no guarantee for convergence, and there is no guarantee that a minimum exists for a given partial derivative.

Generally, when this method does converge, it converges slowly. However, in many practical applications, this method does converge and converges quickly.

Relaxation

I-1   Dependent variable z

I-2   Independent variables $\vec{x}$

I-3   Differentiable function $f(\vec{x})$ relating $z = f(\vec{x})$

I-4   Initial point $\vec{x}^0$

I-5   Uncertainty tolerance ε

S-1   Find the partial derivatives $\dfrac{\partial f(\vec{x})}{\partial x_k}$

S-2   Find $\dfrac{\partial f(\vec{x})}{\partial x_k}$

S-3   Choose a random ordering for minimizing each component

S-4
Find the vector $\left(x_1^n, x_2^n, \ldots, x_l^{n+1}, \ldots, x_k^n\right)$ where

$$\frac{\partial f\left(\vec{x} = \left(x_1^n, x_2^n, \ldots, x_l^{n+1}, \ldots, x_k^n\right)\right)}{\partial x_k} = 0$$

and l is the first component to minimize.

S-5
Find the vector $\left(x_1^n, x_2^n, \ldots, x_m^{n+1}, \ldots, x_l^{n+1}, \ldots, x_k^n\right)$ where

$$\frac{\partial f\left(\vec{x} = \left(x_1^n, x_2^n, \ldots, x_m^{n+1}, \ldots, x_l^{n+1}, \ldots, x_k^n\right)\right)}{\partial x_k} = 0$$

and m is the second component to minimize.

S-6   Repeat the previous step for each component in the randomly chosen order to get the vector for the next iteration $\vec{x}^{n+1} = \left(x_1^{n+1}, x_2^{n+1}, \ldots, x_k^{n+1}\right)$.

S-7   Repeat from S-1 until $\left|\vec{x}^{n+1} - \vec{x}^n\right| < \varepsilon$

O-1 $\vec{x}^{n+1}$ value of a stationary point for the objective function

Algorithm XXII: Relaxation may be used to find a stationary point for a multidimensional objective function.

As an example, examine the objective function

$$f(x, y) = -(x - .5)(x + .5)y^2 - (y - .5)(y + .5)x^2 \qquad 2.114$$

The partial derivatives are

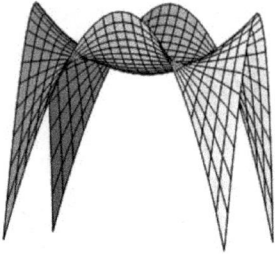

Figure 36: Objective function for the example of the relaxation method.

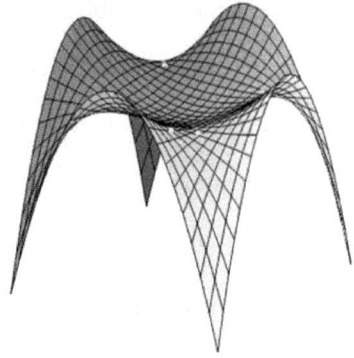

Figure 37: A single iteration of the relaxation method finds all four saddle points.

$$\frac{\partial f(x,y)}{\partial x} = \frac{1}{2}y - 4x^2y \qquad\qquad 2.115$$

$$\frac{\partial f(x,y)}{\partial y} = \frac{1}{2}x - 4y^2x \qquad\qquad 2.116$$

The stationary points may be found from the system of equations

$$\frac{1}{2}y - 4x^2y = 0 \qquad\qquad 2.117$$

$$\frac{1}{2}x - 4y^2x = 0 \qquad\qquad 2.118$$

There is a stationary point at $(x,y) = (0,0)$. Aside from this point, we may divide the equations by x and y:

$$\frac{1}{2} - 4x^2 = 0 \qquad\qquad 2.119$$

$$\frac{1}{2} - 4y^2 = 0 \qquad\qquad 2.120$$

This system has the four solutions

$$(x,y) = \begin{cases} \left(\frac{1}{\sqrt{8}}, \frac{1}{\sqrt{8}}\right) \\ \left(\frac{1}{\sqrt{8}}, -\frac{1}{\sqrt{8}}\right) \\ \left(-\frac{1}{\sqrt{8}}, \frac{1}{\sqrt{8}}\right) \\ \left(-\frac{1}{\sqrt{8}}, -\frac{1}{\sqrt{8}}\right) \end{cases} \qquad\qquad 2.121$$

The stationary point at $(0,0)$ is a maximum, while the points at $\left(\pm\frac{1}{\sqrt{8}}, \pm\frac{1}{\sqrt{8}}\right)$ are all saddle points.

If we apply the relaxation method starting from $(.5, -.25)$ we have:

$$\left.\frac{\partial f(x,y)}{\partial x}\right|_{y=-.25} = -\frac{1}{8} + x^2 = 0 \qquad\qquad 2.122$$

Solving this for x,

$$x = \pm\frac{1}{\sqrt{8}} \qquad\qquad 2.123$$

Applying to y,

$$\left.\frac{\partial f(x,y)}{\partial y}\right|_{x=\pm\frac{1}{\sqrt{8}}} = \pm\frac{1}{2\sqrt{8}} \mp \frac{4}{\sqrt{8}}y^2 = 0 \qquad\qquad 2.124$$

Solving for y,

$$y = \pm\frac{1}{\sqrt{8}} \qquad\qquad 2.125$$

In one iteration we have found all four saddle point solutions.

However, we did not find the maximum at $(0,0)$. In fact, the relaxation method will never find this solution in this case. This demonstrates one of the drawbacks to the relaxation method. There is no guarantee that we will find a particular stationary point no matter what point we choose to start from. In this case, the only way the relaxation method will find the stationary point at $(0,0)$ is if we begin exactly at $\vec{x}^0 = (0,0)$. Any other starting point leads immediately to the four saddle points at $\left(\pm\frac{1}{\sqrt{8}}, \pm\frac{1}{\sqrt{8}}\right)$.

The advantage to the relaxation method is that it effectively changes a n-dimensional multivariate optimization problem into n 1-dimensional optimization problems. This allows us to use the univariate optimization techniques discuss previously to solve the multivariate optimization problem.

Although we are not guaranteed to find every solution, in many cases the relaxation method is able to quickly find solutions. Moreover, we do not need to compute the partial derivatives. Relaxation creates n 1-dimensional optimization problems from an n-dimensional problem. If we apply a 1-dimensional optimization technique that does not require computation of the derivative, we do not need to compute the derivative for the relaxation method.

2.18   Evolutionary Algorithms

Evolutionary algorithms are a class of algorithms that use stochastic methods to iteratively search a space for optimal solutions. Many of these algorithms receive their inspiration from nature as a guide for methods to search for solutions.

Generally, evolutionary algorithms begin with some initial set of seed points. These are typically points chosen at random from the search space. The objective function is evaluated at each of these points. Next, a new set of points is chosen based in part from the original set. The objective function is evaluated at each of the new points. Finally, an evolutionary decision is made to determine which of the old points and new points move on to the next iteration of the algorithm. The entire process is repeated iteratively.

There are many different versions of evolutionary algorithms. This section presents an overview of the general method and a brief introduction to some of the common evolutionary algorithms. The algorithm presented here is a general template for evolutionary algorithms and is not intended for implementation in and of itself. Rather, this template provides a basic understanding of the approach that evolutionary algorithms use for optimizing objective functions.

Evolutionary algorithms may be used with objective functions that have number of variables. However, these algorithms are best suited for objective functions with many variables where other techniques prove too difficult.

Evolutionary Algorithms

I-1	Dependent variable z
I-2	Independent variables $\vec{x}$
I-3	Objective function $f(\vec{x})$ relating $z = f(\vec{x})$
I-4	N initial points $\vec{x}^0{}_n$
S-1	Evaluate the objective function at each of the N points $\vec{x}^k{}_n$
S-2	Identify a set of M test points based on $\vec{x}^k{}_n$
S-3	Evaluate the objective function at each of the M test points
S-4	Use the evolutionary method to select L points for the next iteration
S-5	Repeat from S-1 as many times as desired
O-1	Each $\vec{x}^{n+1}{}_l$ represents a potential optima for the objective function

Algorithm XXIII: Template for the method commonly used by evolutionary algorithms.

The following describes some of the common evolutionary algorithms. There are many such algorithms, and this list is meant as a representative sample.

Genetic Algorithms – Genetic algorithms are based on the concept of sexual reproduction in cells. Here, a chromosome represents a state for the system (a point in the search space or a value $\vec{x}$ for the objective function). The chromosome is made up of individual genes. Different gene sequences represent different states. Two chromosomes are randomly selected from the pool where the selection is weighted by fitness (value of the objective function). The chromosomes are lined up, and a random point is selected for crossover. The second part of one chromosome is copied to the first and vice versa. Pairs are selected and new chromosomes generated. The fitness is computed for each of the resulting chromosomes, and those with the highest fitness are chosen to seed the next generation.

Differential Evolution – Differential evolution begins with a set of initial vectors in the search space. For each vector in the population, randomly choose three other distinct vectors. A new vector is created based on these by replacing one of the components with a weighted average of the original vector and the difference between the other two vectors. If the resulting vector has better fitness than the original vector from the population, the vector is replaced. Otherwise we move onto the next vector in the population. This process is repeated iteratively for each generation.

Particle Swarm Optimization – This technique begins with a set of randomly chosen vectors in the search space. The fitness for each is computed, and the global best fitness is tracked. Each vector is associated with a random velocity. For each iteration, each vectors velocity and position is updated. The velocity is updated using a function of the current position, velocity, a random vector, and the best fitness vector seen to this point. A new fitness is computed and compared with the best fitness thus far for the particular vector in question. If the new fitness is better, the vector is updated to the value of the new fitness. In either case, the velocity is updated for each vector.

CMA-ES – The Covariance Matrix Adaptation Evolution Strategy (CMA-ES) is an approach that begins with an initial vector and test vectors are generated through a probability distribution based on the initial vector. The probability distribution is a multinomial distribution and the covariance matrix for the distribution is used to modify the search process. Based on the evaluation of the objective function at the test vectors, the covariance matrix is updated, and the initial vector is set to the best value of the objective function found thus far. The process repeats itself again for the next generation, and the process is repeated iteratively.

2.19   Genetic Algorithms

Genetic Algorithms are optimization algorithms modeled on the DNA reproduction/replication. In computer terms, the DNA represents some system state (the value of x to test). This process attempts to capture the evolutionary improvement aspects of how organisms evolve as a species.

Just as DNA is made up of chromosomes, genetic algorithms use a chromosome concept. In a genetic algorithm, a chromosome is a sequence (vector) of genes. Each gene may have a plurality of different possible values, and a chromosome is a choice of value for each gene in the sequence.

The chromosome vector is used to identify a point in the search space. A unique sequence of genes maps to a unique point in the search space. However, we may have two different sequences map to the same point in the search space.

The algorithm begins with a set of random chromosomes. The fitness of a chromosome is the value of the objective function based on the point in the search space corresponding to the chromosome. Next, pairs of chromosomes are randomly selected from the population, but the selection process is weighted by the fitness. Higher fitness translates to a higher probability of selecting a chromosome.

Each pairs selected should contain two different chromosomes. However, once we select a pair, we put these back in the population before selecting a new pair. Thus, the same chromosome may appear in multiple pairs. In fact, it is possible that the same pair of chromosomes may appear twice in the list of pairs selected.

Then we perform a crossover or mutation. For a crossover, we swap and combine parts of the DNA strands to form two new strands. For a mutation, we randomly select one gene on each chromosome in the pair and randomly change the gene value to another possible value. In either case, the two resulting new chromosomes are added to the next generation.

This process is repeated for another pair of chromosomes. A new pair of chromosomes is selected, crossover / mutation is chosen, and two new chromosomes are generated. We continue to select pairs of chromosomes and put them into the next generation until we have the desired number of chromosomes for the next generation.

Once the desired number of chromosomes is achieved for the next generation, the fitness for each is computed. If the best fitness of this set is not at least as good as the best fitness from the previous generation, we remove one chromosome from the next generation and add the best chromosome from the previous generation.

The entire process is repeated for each generation. At each generation, the best fitness must increase or at least stay the same. By repeating the process iteratively, we continually increase the fitness and achieve an increasingly improving performance.

Genetic Algorithms

I-1	Dependent variable z
I-2	Independent variables $\vec{x}$
I-3	Identify the genes and a method for mapping gene sequences to vectors in the search space
I-4	Objective function $f(\vec{x})$ relating $z = f(\vec{x})$ where $\vec{x}$ is a sequence of genes (a chromosome)
I-5	N initial genes $\vec{x}^0{}_n$
I-6	A weighted random selection method for randomly choosing a chromosome from the population pool, where the weight of the random selection is dependent on the value of the objective function
S-1	Evaluate the objective function at each of the N points $\vec{x}^k{}_n$
S-2	Randomly select two chromosomes from the population pool. The random selection process is weighted by the value of the objective function.
S-3	Choose crossover or mutation. This may be a weighted choice or simply a 50-50 choice. Variants of the algorithm include only using crossover or only using mutation.
S-4	Perform crossover or mutation on the two chromosomes randomly selected.
S-5	Place the two new chromosomes into the population pool for the next generation.
S-6	Repeat from S-2 until the desired number of chromosomes for the next generation population pool is achieved.
S-7	Evaluate the objective function for each chromosome in the next generation population pool.

S-8	If the best fitness for the next generation population pool is less than the best fitness from the current population pool, randomly select one of the chromosomes from the next generation and replace it with the best chromosome from the current generation.
S-9	Repeat from S-2 as many times as desired.
O-1	The best fitness from the current generation is potential optimum. The corresponding chromosome maps to the point in the state space where the optimum occurs.

Algorithm XXIV: Genetic algorithms are evolutionary algorithms used to find an optimum for the objective function.

Crossover and mutation are used to generate new states from the previous states. These methods work as follows:

Crossover – Start with two chromosomes (state vectors) x_1 and x_2. Choose a gene (component) within the states. Swap part of x_1 with the opposite part of x_2. Generate another state by combining the other part of x_1 with x_2.

Mutation – Start with a single chromosome (state vector). Pick a gene within the chromosome (one component of the state vector). Randomly change the value of this part to another acceptable value.

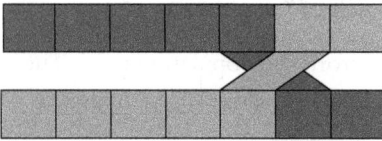

Figure 38: Crossover between two states.   Figure 39: Mutation of a state.

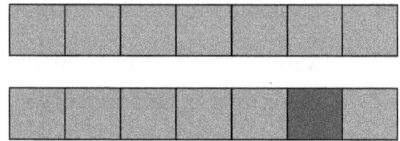

This basic process often converges quickly on optimal solutions. Moreover, the process is fairly simple to implement in software systems.

There are many variations to this process. It should be noted that Genetic Algorithms can be based entirely on mutation with no crossover whatsoever. This may be an effective technique when working with systems where two different states cannot necessarily be represented with a uniform genetic representation.

As a simple example, take the objective function

$$f(x) = -x(x-1) \qquad\qquad 2.126$$

To encode the chromosomes, let each gene have the value either 0 or 1. Let a chromosome be a sequence of genes, and interpret this sequence as a binary number. If we use sixteen genes per chromosome, then we have a sixteen bit number. The minimum occurs when all genes have the value zero, and this corresponds to the number zero. The maximum occurs when all genes have the value one, and this corresponds to the integer $2^{16} - 1 = 65535$. The search space is the range [0,1]. To map the chromosome values to the search space, we divide the binary value from the chromosome by the max value 65535.

With ten chromosomes in the population pool, we get the results shown in Figure 40-Figure 43. Initially, the population pool is distributed randomly over the search space as shown in Figure 40. With each generation, the population continues to cluster near the global maximum at $x = .5$. With only three generations, 80% of the chromosomes are near the global maximum.

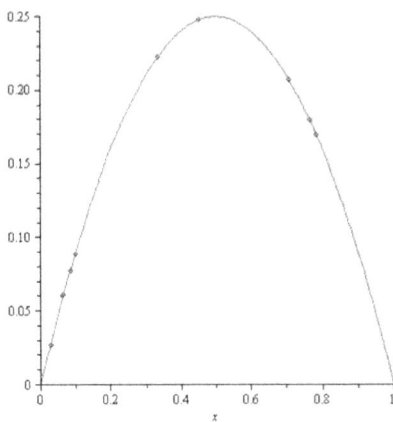

Figure 40: Genetic algorithm at the initial state.

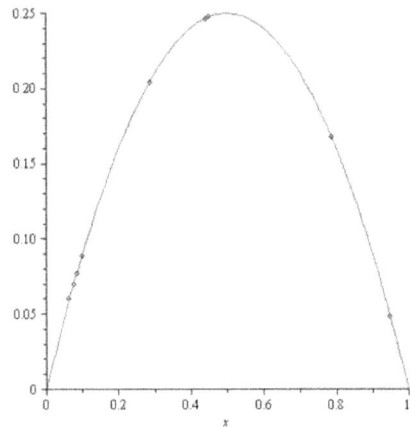

Figure 41: Genetic algorithm after one generation.

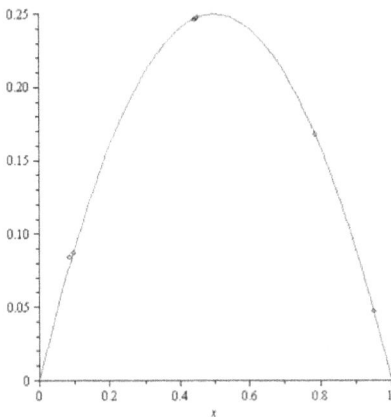

Figure 42: Genetic algorithm after two generations.

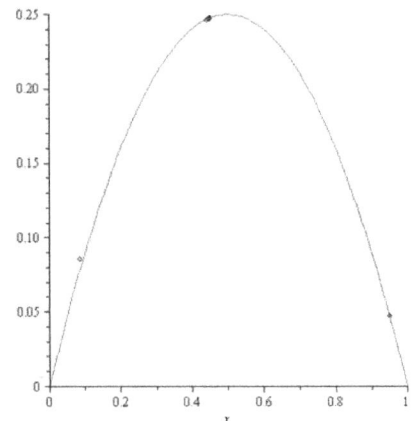

Figure 43: Genetic algorithm after three generations.

2.20   Quantum Particle Search

Quantum particle search uses the form of the Feynman path integral as a method of searching. The Feynman path integral determines how quantum particles move in space. The probability that a particle takes a given path is given by

$$exp\left(\frac{i}{\hbar}\int_a^b \mathcal{L}dt\right)$$

2.127

where $\mathcal{L}$ is the Lagrangian of the physical system. Quantum particle search starts with the state of a particle, then generates a random state based on this state with a Laplace probability density. The Laplace density is given by

$$P(x) = \frac{\alpha}{2}exp(-\alpha|x|)$$

2.128

Thus, given an initial state vector with components x_i, the test vector is

$$\vec{t} = \vec{x} + \vec{r}$$

2.129

where the components of $\vec{r}$ are random variables according to the Laplace density.

The algorithm proceeds by examining each state vector in the population, computing a test vector based on the state vector, and computing the value of the objective function at the test vector. If $f(\vec{t})$ is better than $f(\vec{x})$, then $\vec{t}$ replaces $\vec{x}$ in the population.

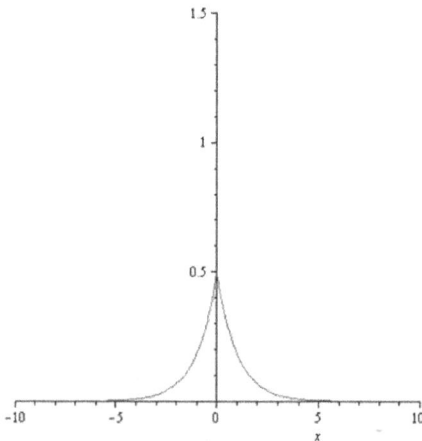

Figure 44: Example of the Laplace distribution.   Figure 45: Laplace distribution at $\alpha = 1, 2$, and 3. As α increases, the distribution has a higher peak and smaller variance.

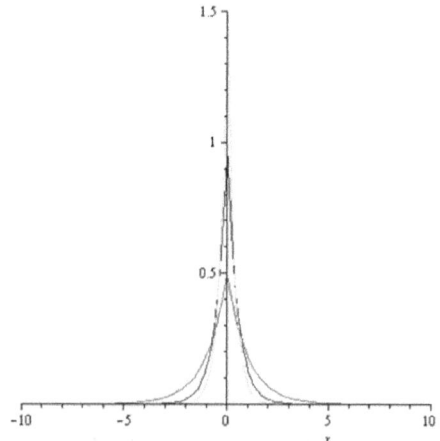

Random variables under a Laplace density may be computed from uniform deviates. If we have a random variable u on the range $[0,1)$ taken from a uniform distribution, a Laplace distributed variable may be constructed as

$$v = -sign\left(u - \frac{1}{2}\right) ln\left(1 - 2\left|u - \frac{1}{2}\right|\right)$$

2.130

The method may be modified to use a value of α that becomes smaller as the number of iterations increases. This provides a large search range for the initial iterations, then decreases the variance as the number of iterations increases. Thus, at first we search widely across the space, and then as the states converge to an optimum, we test only nearby states. This allows the algorithm an opportunity to escape local optima in the beginning, and then converge to the global optimum as the number of iterations increases.

Quantum Particle Search

I-1   Dependent variable z

I-2   Independent variables $\vec{x}$

I-3   Objective function $f(\vec{x})$ relating $z = f(\vec{x})$

I-4 N initial states x_i

I-5   Value α used in the Laplace probability density

S-1   Iterate over each state $\vec{x}_i$

S-2   Compute $f(\vec{x}_i)$ for each state

S-3   Generate a random vector $\vec{r}$ where the components of $\vec{r}$ are generated according to the Laplace density

$$P(x) = \frac{\alpha}{2} exp(-\alpha|x|)$$

S-4   Compute

$$\vec{t} = \vec{x} + \vec{r}$$

S-5   If $f(\vec{t})$ is better than $f(\vec{x}_i)$, then replace $\vec{x}_i$ with $\vec{t}$ and process the next state

O-1   The best value of $f(\vec{x}_i)$ is the potential optimum.

Algorithm XXV: Quantum particle search is useful for discovering an optimum for the objective function.

2.21   Differential Evolution

Differential evolution is an optimization technique originally designed to extend genetic programming concepts to use continuous variables. The differential evolution method provides a reliable convergence over a wide range of problems.

The method begins with a set of randomly initialized state vectors. A state vector is a vector in the n-dimensional problem space and is represented as a n-component vector of real numbers. A population size is chosen (N), and the initial vector values are randomly selected within the problem space.

The algorithm proceeds iteratively. During an iteration, each state vector is considered one at a time. When considering a state vector, three other state vectors are chosen at random. These three state vectors must be distinct from each other and distinct from the state vector under consideration.

One random vector component is chosen. Each vector component is considered individually. A new state vector is constructed as follows:

For the randomly chosen component, set the component of the new vector to the value

$$v_r = a_r + F(b_r - c_r) \qquad\qquad 2.131$$

where a, b, and c are the three other vectors randomly selected, r is the randomly selected component, and F is a control variable set by the user.

For each other vector component, crate a random variable $R \in (0,1)$. The user sets a crossover probability CP. If $R < CP$ then compute the component using the equation above. Otherwise, use the value of this component from the state vector under consideration.

After constructing the new state vector, we compute the value of the objective function for this state. If the value for the constructed vector is better than the original state vector under consideration, in the next generation we replace this vector with the newly constructed vector.

This process is repeated iteratively with a new set of state vectors created at each iteration. Since we only replace a state vector with a better valued state, we are guaranteed that the best value for each iteration is at least as good as the best from the previous iteration.

The differential evolution method is used effectively for a wide variety of problems and has demonstrated robust capability for effective convergence. The method is easy to implement as a software solution and can be constructed flexibly to allow application to nearly any vector valued function.

Differential Evolution

I-1   Dependent variable z

I-2   Independent variables $\vec{x}$

I-3   Objective function $f(\vec{x})$ relating $z = f(\vec{x})$

I-4 N initial states $x^k{}_n$ (upper index is the state number, lower index designates the vector component)

I-5   Value for F

I-6   Value for CP

S-1   Evaluate the objective function at each of the N points $\vec{x}^k{}_n$

S-2   Iterate over each state vector $x^k{}_i$

S-3   Choose three distinct random states different from each other and $x^k{}_i$. Designate these as $x^a{}_i, x^b{}_i$, and $x^c{}_i$

S-4   Select a random vector component r

S-5   Construct a new state y_i from $x^k{}_i$ and $x^a{}_i, x^b{}_i$, and $x^c{}_i$ by examining each component:

S-6   Set $y_r = x^a{}_r + F\left(x^b{}_i - x^c{}_i\right)$

S-7   For the components $j \neq r$:

Choose a random variable R. If $R < CP$ then set

$$y_j = x^a{}_j + F\left(x^b{}_j - x^c{}_j\right)$$

otherwise set

$$y_j = x^k{}_j$$

S-8   If $f(y_r)$ is better than $f\left(x^k{}_i\right)$, then replace $x^k{}_i$ with y_r in the next generation. Otherwise, use $x^k{}_i$ in the next generation.

S-9   Repeat from S-2 for each state vector.

S-10   Repeat from S-1 for each iteration

O-1   The best fitness from the current generation is potential optimum.

Algorithm XXVI: Differential evolution is a type of evolutionary algorithm used to find an optimum for the objective function.

As an example, examine the objective function

$$f(x) = -x(x - 1) \qquad\qquad 2.132$$

where x is on the range $x \in [0,1]$ and we wish to maximize x.

Figure 46-Figure 49 show application of differential evolution to this one-dimensional problem with 10 states, $F = .5$, and $CP = .25$. A one-dimensional example such as this does not showcase the true utility of the algorithm. However, this example is useful for graphically displaying the iterative results and comparison with other algorithms.

One difficulty encountered with differential evolution and similar algorithms is that the generated test states are not necessarily contained in the operational region specified. In our example, we may generate values of x where $x < 0$ or $x > 1$. This simple constraint is easy to handle by simply setting x to the limit of the operational range.

In larger dimensions this may become an increasingly difficult problem to handle. This is especially true if the operational range is not a simple geometric constraint (a hypercube of allowable values) but is a patchwork region with holes where states are not allowed. If the state constraints are easy to test and identify, we may simply choose to ignore states generated in unallowable regions. However, if the space is highly constrained where the allowable regions are few and far between, it may be difficult to generate appropriate states to test.

Differential evolution, like most of the algorithms we discuss, can be modified in many ways to incorporate these types of constraints. The basic algorithm presented in the table above does not incorporate these methods. As such, when characterizing this algorithm, we describe it as one that does not apply to constrained optimization. We mean this in the sense that the base algorithm does not incorporate constraints. However, we recognize that there are many methods for incorporating constraints.

One of the key benefits to the differential evolution method is the simplicity of the base algorithm. In many applications, evaluating the objective function is very computationally expensive. In these situations, it is important to have a base algorithm that operates quickly and efficiently. When the base algorithm is complicated itself, the overhead of the operation of the algorithm in conjunction with the computational time required to evaluate the objective function may render an approach computationally useless.

Moreover, each evaluation of the objective function is a direct attempt to improve the fitness. If we were to use an algorithm that relied on taking a derivative at a test point, we would need one evaluation of the objective function to determine the fitness at the point, but then a separate, nearby evaluation of the objective function in order to numerically compute the derivative.

Algorithms such as these require that the objective function is evaluated multiple times in order to operate a single test point. Algorithms that rely on derivatives, especially higher order derivatives, will require multiple evaluations of the objective function in order to process a single test point. Again, in situations where the objective function is computationally expensive to compute, this additional overhead may render the algorithm computationally useless.

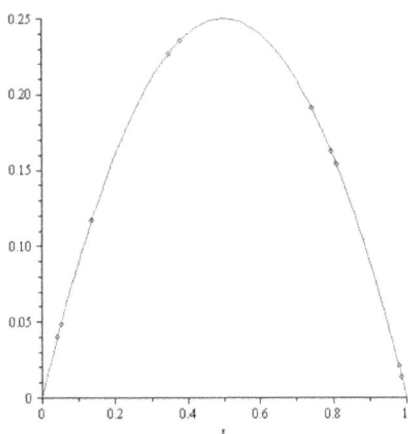

Figure 46: Differential evolution at the initial state.

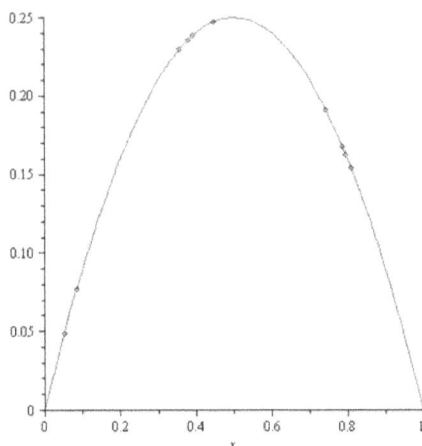

Figure 47: Differential evolution after one generation.

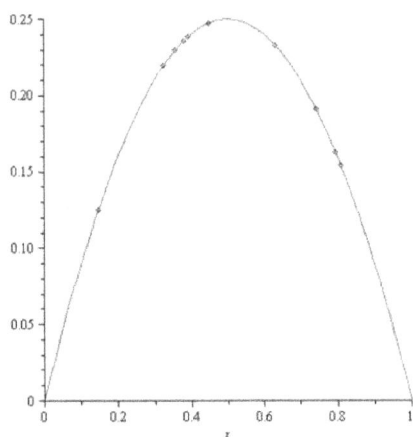

Figure 48: Differential evolution after two generations.

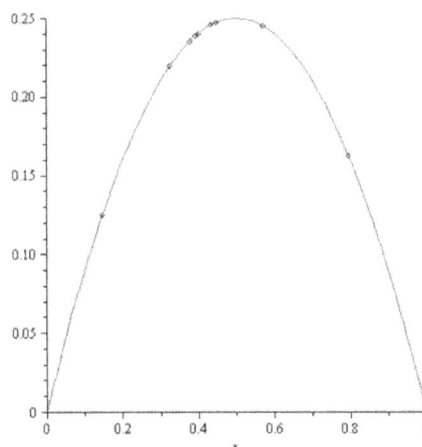

Figure 49: Differential evolution after three generations.

2.22   Particle Swarm Optimization

Particle swarm optimization is another evolutionary algorithm using stochastic processes to discover global optima. This method uses a set of distinct particles and computes best known position for each particle as well as the overall best known position for the swarm (all particles). This information is used to compute a velocity for each particle, which is then used to update the position of the particle to test new points in the search space.

The graphs in Figure 50-Figure 53 show the first few iterations of particle swarm optimization to the example problem used previously.

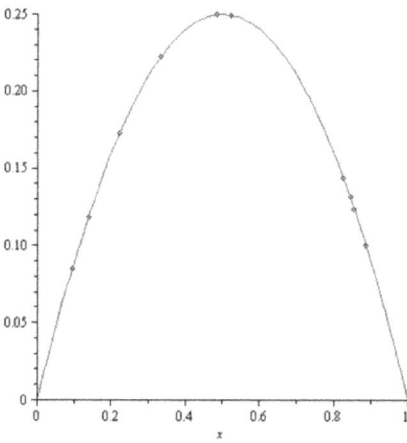

Figure 50: Particle swarm optimization at the initial state.

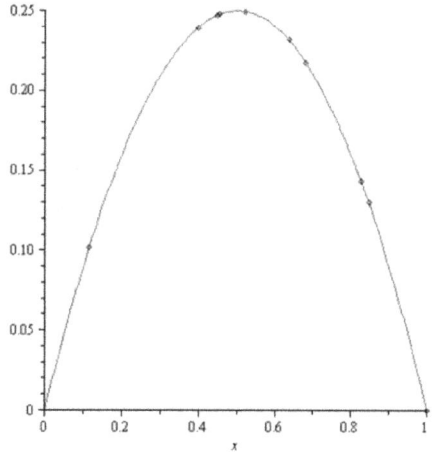

Figure 51: Particle swarm optimization after one generation.

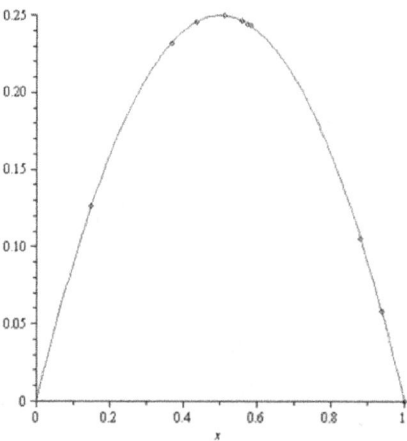

Figure 52: Particle swarm optimization after two generations.

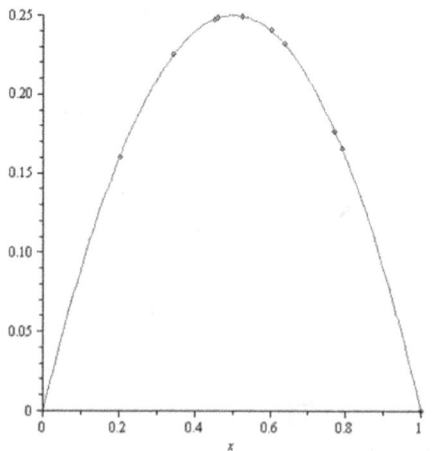

Figure 53: Particle swarm optimization after three generations.

Particle Swarm Optimization

I-1	Dependent variable z
I-2	Independent variables $\vec{x}$
I-3	Objective function $f(\vec{x})$ relating $z = f(\vec{x})$
I-4	N initial states $x^k{}_n$ randomly chosen in the search space
I-5	Initial velocity $v^k{}_n$ randomly chosen for each state
I-6	Value for ω
I-7	Value for ρ
I-8	Value for γ
I-9	Set the best known position for each particle to the initial position $$p^k{}_i = x^k{}_i$$
I-10	Evaluate the objective function at each of the N points $x^k{}_n$
I-11	Set the best swarm position $\vec{s}_i$ to the state vector that has the best value for the objective function
S-1	For each state vector $x^k{}_i$:
S-2	Choose two random variables, R_1 and R_2, on the range $[0,1]$.
S-3	Update the velocity of the state by $$v^k{}_i = \omega v^k{}_i + \rho R_1\left(p^k{}_i - x^k{}_i\right) + \gamma R_2\left(s_i - x^k{}_i\right)$$
S-4	Update the position of the state by $$x^k{}_i = x^k{}_i + v^k{}_i$$
S-5	Compute the value of the objective function for this state $f\left(x^k{}_i\right)$
S-6	If $f\left(x^k{}_i\right)$ is better than $f\left(p^k{}_i\right)$ then set $p^k{}_i = x^k{}_i$
S-7	If $f\left(x^k{}_i\right)$ is better than $f(s_i)$ then set $s_i = x^k{}_i$
S-8	Repeat from S-1 for each state vector.
O-1	s_i is the potential optimum.

Algorithm XXVII: Particle swarm optimization is useful for discovering an optimum for the objective function.

2.23   Covariance Matrix Adaption

Covariance matrix adaption begins with a candidate solution specified as the n-dimensional vector $\vec{x}^0$. Based on this vector, a set of test points are generated using a multinomial normal distribution. The test points are evaluated using the objective function, and this information is used to update the candidate solution. The process is repeated iteratively.

The multinomial normal distribution used is governed by a covariance matrix. The probability density function is given by

$$p(\vec{x}) = \frac{1}{(2\pi)^{k/2}\sqrt{|\Sigma|}} exp\left(-\frac{1}{2}(\vec{x} - \vec{\mu})^T \Sigma^{-1}(\vec{x} - \vec{\mu}) \right) \qquad 2.133$$

The covariance matrix Σ governs the distributional relationship between the different components in $\vec{x}$. In general, the covariance matrix may be expressed as the joint expectation between the variables

$$\Sigma = \begin{bmatrix} \langle(x_1 - \mu_1)(x_1 - \mu_1)\rangle & \langle(x_1 - \mu_1)(x_2 - \mu_2)\rangle & \cdots & \langle(x_1 - \mu_1)(x_n - \mu_n)\rangle \\ \langle(x_2 - \mu_2)(x_1 - \mu_1)\rangle & \langle(x_2 - \mu_2)(x_2 - \mu_2)\rangle & \cdots & \langle(x_2 - \mu_2)(x_n - \mu_n)\rangle \\ \vdots & \vdots & \vdots & \vdots \\ \langle(x_n - \mu_n)(x_1 - \mu_1)\rangle & \langle(x_n - \mu_n)(x_2 - \mu_2)\rangle & \cdots & \langle(x_n - \mu_n)(x_n - \mu_n)\rangle \end{bmatrix} \qquad 2.134$$

or

$$\Sigma = \begin{bmatrix} \sigma_{11} & \sigma_{12} & \cdots & \sigma_{1n} \\ \sigma_{21} & \sigma_{22} & \cdots & \sigma_{2n} \\ \vdots & \vdots & \vdots & \vdots \\ \sigma_{n1} & \sigma_{n2} & \cdots & \sigma_{nn} \end{bmatrix} \qquad 2.135$$

where

$$\sigma_{ij} = \langle(x_i - \mu_i)(x_j - \mu_j)\rangle \qquad 2.136$$

If Σ is a diagonal matrix, then the components are independent of each other and each diagonal element is related to the variance of the distribution for the parameter. For independent variables,

$$\Sigma = \begin{bmatrix} \sigma_{11} & \cdots & 0 \\ \vdots & \ddots & \vdots \\ 0 & \cdots & \sigma_{nn} \end{bmatrix} \qquad 2.137$$

For a one-dimensional problem, the covariance matrix is

$$\Sigma = \sigma_{xx} = \langle(x - \mu)^2\rangle = \sigma^2 \qquad 2.138$$

The probability density is then

$$p(x) = \frac{1}{\sqrt{2\pi}\sigma} exp\left(-\frac{(x-\mu)^2}{2\sigma}\right)$$ 2.139

This is the standard Gaussian distribution. In two dimensions we have,

$$\Sigma = \begin{bmatrix} \sigma_{xx} & \sigma_{xy} \\ \sigma_{yx} & \sigma_{yy} \end{bmatrix}$$ 2.140

The off-diagonal elements are the same because

$$\sigma_{xy} = \langle(x-\mu_x)(y-\mu_y)\rangle = \langle(y-\mu_y)(x-\mu_x)\rangle = \sigma_{yx}$$ 2.141

Thus,

$$\Sigma = \begin{bmatrix} \sigma_{xx} & \sigma_{xy} \\ \sigma_{xy} & \sigma_{yy} \end{bmatrix}$$ 2.142

and

$$|\Sigma| = \sigma_{xx}\sigma_{yy} - \sigma_{xy}{}^2$$ 2.143

The inverse of the covariance matrix is

$$\Sigma^{-1} = \frac{1}{|\Sigma|}\begin{bmatrix} \sigma_{yy} & -\sigma_{xy} \\ -\sigma_{xy} & \sigma_{xx} \end{bmatrix}$$ 2.144

With this, the probability density is

$$p(x,y) = \frac{1}{2\pi\sqrt{|\Sigma|}} exp\left(-\frac{1}{2|\Sigma|}[x-\mu_x, y-\mu_y]\begin{bmatrix} \sigma_{yy} & -\sigma_{xy} \\ -\sigma_{xy} & \sigma_{xx} \end{bmatrix}\begin{bmatrix} x-\mu_x \\ y-\mu_y \end{bmatrix}\right)$$ 2.145

$$= \frac{1}{2\pi\sqrt{|\Sigma|}} exp\left(-\frac{1}{2|\Sigma|}[(x-\mu_x)\sigma_{yy} - (y-\mu_y)\sigma_{xy}, -(x-\mu_x)\sigma_{xy}\right.$$
$$\left. + (y-\mu_y)\sigma_{xx}]\begin{bmatrix} x-\mu_x \\ y-\mu_y \end{bmatrix}\right)$$ 2.146

$$= \frac{1}{2\pi\sqrt{|\Sigma|}} exp\left(-\frac{1}{2|\Sigma|}[(x-\mu_x)^2\sigma_{yy} - 2(x-\mu_x)(y-\mu_y)\sigma_{xy} + (y-\mu_y)^2\sigma_{xx}]\right)$$ 2.147

This may be written in terms of the correlation between the variables

$$\rho = \frac{\sigma_{xy}}{\sigma_x\sigma_y}$$ 2.148

Substituting this into the expression for the probability density,

$$= \frac{1}{2\pi\sqrt{|\Sigma|}} exp\left(-\frac{1}{2|\Sigma|}[(x-\mu_x)^2\sigma_y{}^2 - 2(x-\mu_x)(y-\mu_y)\rho\sigma_x\sigma_y + (y-\mu_y)^2\sigma_x{}^2]\right)$$ 2.149

Note that

$$|\Sigma| = \sigma_x{}^2\sigma_y{}^2 - \rho^2\sigma_x{}^2\sigma_y{}^2 = \sigma_x{}^2\sigma_y{}^2(1-\rho^2)$$ 2.150

With this the probability density is

$$= \frac{1}{2\pi\sigma_x\sigma_y\sqrt{(1-\rho^2)}} exp\left(-\frac{1}{2\sigma_x^2\sigma_y^2(1-\rho^2)}\left[(x-\mu_x)^2\sigma_y^2 - 2(x-\mu_x)(y-\mu_y)\rho\sigma_x\sigma_y + (y-\mu_y)^2\sigma_x^2\right]\right) \qquad 2.151$$

$$= \frac{1}{2\pi\sigma_x\sigma_y\sqrt{(1-\rho^2)}} exp\left(-\frac{1}{2(1-\rho^2)}\left[\frac{(x-\mu_x)^2}{\sigma_x^2} - 2\rho\frac{(x-\mu_x)(y-\mu_y)}{\sigma_x\sigma_y} + \frac{(y-\mu_y)^2}{\sigma_x^2}\right]\right) \qquad 2.152$$

This process may be extended to higher dimensions to find expressions for the probability density in terms of the elements of the covariance matrix.

With the probability density specified, the algorithm proceeds as follows. First, an initial point is selected as the mean for the distribution $\vec{\mu}^0$. The covariance matrix is initially set to the identity matrix $\Sigma^0 = I$. Also, an initial step size is selected $s^0 > 0$. Finally, we use a path vector for the step size and a path vector for the covariance matrix. These are designated $\vec{p}_s$ and $\vec{p}_\Sigma$ respectively and both are initialized to zero.

The process begins by generating R random state vectors based on the probability density. A method for generating variables from this distribution is provided in Appendix B: Gaussian Sampling. Let $\mathcal{N}(\vec{\mu}, \Sigma)$ represent a sample vector from the multivariate Gaussian distribution with mean $\vec{\mu}$ and covariance matrix Σ. The randomly generated state vectors have the form

$$\vec{r} = \vec{\mu} + s\mathcal{N}(\vec{\mu}, \Sigma) \qquad 2.153$$

Next we compute the objective function for each of the random vectors generated. The results are rank ordered from best to worst so that the vector $\vec{r}_1$ corresponds to the random vector that has the best value for the objective function, $\vec{r}_2$ the next best, etc.

Then using the best k of the vectors, we choose a weighting scheme such that

$$w_1 \geq w_2 \geq \cdots \geq w_k > 0 \qquad 2.154$$

such that

$$\beta = \sum_{i=1}^{k} w_i^2 \approx \frac{4}{N} \qquad 2.155$$

where N is the number of dimensions in the state space.

The value for the distribution mean is updated as

$$\vec{\mu}^1 = \sum_{i=1}^{k} w_i\vec{r}_i \qquad 2.156$$

The search paths are updated as

$$\vec{p}_\Sigma^{\,1} = \alpha \vec{p}_\Sigma^{\,0} + \sqrt{1 - \alpha^2} \sqrt{\beta} \Sigma^{-1/2} \frac{\vec{\mu}^1 - \vec{\mu}^0}{s} \qquad 2.157$$

and

$$\vec{p}_s^{\,1} = \alpha' \vec{p}_s^{\,0} + \mathbf{1}_{\left[0, \frac{3\sqrt{N}}{2}\right]} (|\vec{p}_\Sigma^{\,1}|) \sqrt{1 - \alpha'^2} \sqrt{\beta} \Sigma^{-1/2} \frac{\vec{\mu}^1 - \vec{\mu}^0}{s} \qquad 2.158$$

Finally, the covariance matrix is updated as

$$\Sigma^1 = \gamma \Sigma^0 + \delta \vec{p}_s^{\,1} \cdot \vec{p}_s^{\,1} + \varepsilon \sum_{i=1}^{k} w_i \left(\frac{\vec{r}_i - \vec{\mu}^1}{s} \right) \left(\frac{\vec{r}_i - \vec{\mu}^1}{s} \right)^T \qquad 2.159$$

Where the Greek letters α-ε are control parameters set by the user.

The covariance matrix adaption method is more complicated than many of the other methods examined thus far. There are several control parameters that may be set by the user as well.

This provides the method with a great deal of flexibility to tune the method to work well with a specific problem. However, in many cases this is a self-defeating proposition. Often we desire to find the optimum for a single problem. To tune a method we must solve the problem using a set of parameters, then modify the parameters and solve the problem again. Each time we solve the problem we use a different set of parameters with the intention to discover the optimal values for the parameters.

However, our goal is not to find the optimal set of parameters, our goal is to solve the problem. In tuning the optimization method, we must repeatedly solve the problem. This defeats the utility in finding the best values for the parameters.

There are many possible ranges of values for the parameters that have been noted to work well for this method. However, there is no strict set of guidelines applicable to approaching general optimization methods with this algorithm. In many cases, other algorithms are shown to perform superior to this method without the need to specify as many control parameters.

Moreover, the matrix and vector manipulations required to use this method may limit the utility of the algorithm. For example, we require the inverse of the covariance matrix for the probability density. Matrix manipulations such as this are often computationally expensive. If the optimization problem at hand has an objective function that is very difficult to compute, then the time required for the overhead of this algorithm may be insignificant in comparison. In other cases this method may have too much overhead to be of use.

2.24   Amoeba Method

The amoeba method is also known as the Nelder-Mead method or simplex method. The method uses a simplex which is a set of $n + 1$ vertices in an n-dimentional search space. In two dimensions this forms a triangle, and in three dimensions a tetrahedron.

The method starts with a randomly selected set of vertices. The objective function is computed for each vertex, and the vertices are ordered from best to worst. Let $\vec{v}_i$ be the ith vertex in the order. The mean is computed as

$$\vec{\mu} = \frac{1}{N}\sum_{i=1}^{N} \vec{v}_i \qquad\qquad 2.160$$

Note that we do not include the last (worst) vector when computing the mean. Next we compute the new state vector

$$\vec{x}_s = \vec{\mu} + \alpha(\vec{\mu} - \vec{v}_{N+1}) \qquad\qquad 2.161$$

where α is a control parameter set by the user.

The value of the objective function is computed for the new state vector. If the new state vector is better than the second worst vertex, but not better than the best vertex, we replace the worst vertex with the new state vector. If the new state vector is better than the best vertex, then we compute the expansion vector

$$\vec{x}_e = \vec{\mu} + r(\vec{\mu} - \vec{v}_{N+1}) \qquad\qquad 2.162$$

where r is a control variable. Next, compute the value of the objective function for the expansion vector. If the expansion vector is better than the new state vector, then we replace the worst vertex with the expansion vector. Otherwise, we replace the worst vertex with the new state vector.

Finally, if the new state vector is worse than the worst vertex, then we compute the contraction vector

$$\vec{x}_c = \vec{v}_{N+1} + \rho(\vec{\mu} - \vec{v}_{N+1}) \qquad\qquad 2.163$$

where ρ is a control variable. Compute the objective function for the contracted vector. If the contracted vector is better than the worst vertex, then replace the worst vertex with the contracted vector. Otherwise, replace all vertices but the best vertex using

$$\vec{v}_i = \vec{v}_1 + \sigma(\vec{v}_i - \vec{v}_1) \qquad\qquad 2.164$$

Amoeba Method

I-1   Dependent variable z

I-2   Independent variables $\vec{x}$

I-3   Objective function $f(\vec{x})$ relating $z = f(\vec{x})$

I-4 $N + 1$ initial vertices $\vec{v}_i$

I-5   Value for α

I-6   Value for r

I-7   Value for ρ

I-8   Value for σ

S-1   Compute $f(\vec{v}_i)$ for each vertex

S-2   Order the vertices according to their corresponding values for the objective functions, with the best value being the first vertex, second best as the second vertex, etc.

S-3   Compute

$$\vec{\mu} = \frac{1}{N} \sum_{i=1}^{N} \vec{v}_i$$

Note the sum does not include the worst vertex.

S-4   Compute

$$\vec{x}_s = \vec{\mu} + \alpha(\vec{\mu} - \vec{v}_{N+1})$$

and the corresponding value for the objective function.

S-5   If $f(\vec{x}_s)$ is better than the second worst vertex, but not better than the best vertex, replace the worst vertex with $\vec{x}_s$ and repeat from S-1.

     If $f(\vec{x}_s)$ is better than the best vertex, then compute

$$\vec{x}_e = \vec{\mu} + r(\vec{\mu} - \vec{v}_{N+1})$$

S-6   If $f(\vec{x}_e)$ is better than $f(\vec{x}_s)$, then replace the worst vertex with $\vec{x}_e$ and repeat from S-1.

     Otherwise, replace the worst vertex with $\vec{x}_s$ and repeat from S-1.

     If $f(\vec{x}_s)$ is not better than the second worst vertex, compute

S-7 $$\vec{x}_c = \vec{v}_{N+1} + \rho(\vec{\mu} - \vec{v}_{N+1})$$

     If $f(\vec{x}_c)$ is better than the worst vertex, then replace the worst vertex with $\vec{x}_c$ and repeat from S-1.

     Replace all vertices except the best vertex using the formula

S-8 $$\vec{v}_i = \vec{v}_1 + \sigma(\vec{v}_i - \vec{v}_1)$$

     Repeat from S-1.

O-1 $\vec{v}_1$ is the potential optimum.

Algorithm XXVIII: The amoeba method is useful for discovering an optimum for the objective function.

2.25   Simulated Annealing

Simulated annealing is an optimization approach that mimics the process of annealing in metallurgy. Simulated annealing allows transitions from a given state to a worse state under certain conditions. This allows the algorithm a change to move out of local minima or maxima and move toward the global optima.

Simulated annealing begins with a set of randomly chosen state vectors $\vec{x}_i^{\,0}$. Each iteration produces a new generation of state vectors based on the prior generation. Each vector in the current generation gives rise to a vector in the next generation.

For a given state vector $\vec{x}_i$, a new test vector is chosen using a random selection method. Let $\vec{r}_i$ designate the random vector associated with the state vector $\vec{x}_i$. The value of the objective function is computed for $\vec{r}_i$. If $f(\vec{r}_i)$ is better than $f(\vec{r}_i)$, then $\vec{r}_i$ replaces $\vec{x}_i$ in the next generation. Otherwise, compute

$$\Delta = -|f(\vec{r}_i) - f(\vec{r}_i)| \qquad\qquad 2.165$$

Based on this, a probability of transitioning is computed. The function governing the probability depends on the value of Δ and the 'temperature' T. One such function is

$$P(\Delta, T) = e^{-\Delta/T} \qquad\qquad 2.166$$

When the temperature is high, the exponential is nearly unity for every value of Δ. This allows the state to move freely. As the temperature declines, the exponential gets closer and closer to zero. This 'freezes' the states in place and only allows transitions to better states.

Typically, the temperature is decreased with each iteration. Thus, the temperature starts high, then decreases over time. At first, the states are allowed to transition anywhere within the state space. Then as the number of iterations increases, the states are frozen in place and eventually are only able to move to better states.

The simulated annealing method is simple to implement and does not require a large computational overhead to process the algorithm. This makes the algorithm a useful technique for many optimization problems.

In addition, by controlling the decline in temperature, the algorithm has opportunity to climb out of local minima or maxima. This helps to prevent the algorithm becoming stuck near a local optima and increases the chance for convergence to the global optima.

There are many possible techniques for computing a random vector based on a given state vector. A simple technique is to use a Gaussian or exponential distributed random variable centered at the state vector. By controlling the variance of the distribution, we can control just how close the random vector is to the original vector.

In addition, there are many possible methods for computing the transition probability. The characteristic of this method is that the transition probability allows a non-zero chance to transition to a less desirable state, and this probability depends on the difference in the value of the objective function between the states and the current temperature.

Simulated Annealing

I-1   Dependent variable z

I-2   Independent variables $\vec{x}$

I-3   Objective function $f(\vec{x})$ relating $z = f(\vec{x})$

I-4 N initial states x_i

I-5   Schedule for the value of the temperature at each iteration.

S-1   Iterate over each state $\vec{x}_i$

S-2   Compute $f(\vec{x}_i)$ for each state

S-3   Generate a random test state, $\vec{r}_i$, based on $\vec{x}_i$

S-4   Compute
$$\Delta = -|f(\vec{r}_i) - f(\vec{r}_i)|$$

S-5   If $f(\vec{r}_i)$ is better than $f(\vec{x}_i)$, then replace $\vec{x}_i$ with $\vec{r}_i$ and process the next state

S-6   Otherwise compute the transition probability
$$P(\Delta, T) = e^{-\Delta/T}$$

S-7   Select a random variable $R \in [0,1)$

S-8   If $R < P(\Delta, T)$ then replace $\vec{x}_i$ with $\vec{r}_i$. Process the next iteration.

O-1   The best value of $f(\vec{x}_i)$ is the potential optimum.

Algorithm XXIX: Simulated annealing is useful for discovering an optimum for the objective function.

2.26   Quadratic Vector Interpolation

Quadratic interpolation begins with a set of n state vectors. A pair of vectors is chosen and a third vector is chosen on the line between these points. Based on these three vectors, a quadratic function is interpolated along the line through the three points. The location of the optimum for the quadratic is computed, and the objective function evaluated at this point. We begin by selecting two state vectors. Designate these as $\vec{x}_1$ and $\vec{x}_2$. Set

$$\vec{d} = \vec{x}_i - \vec{x}_j \qquad\qquad 2.167$$

In another variant, we select four vectors and replace $\vec{x}_j$ with the mean of three of the vectors:

$$\vec{M} = \frac{1}{3}(\vec{x}_1 + \vec{x}_2 + \vec{x}_3) \qquad\qquad 2.168$$

Next, compute a random value r on a uniform distribution, and calculate

$$\vec{\sigma} = \vec{x}_j + r\vec{d} \qquad\qquad 2.169$$

A quadratic function may be interpolated based on these three points. The quadratic has the form

$$F(t) = a + bt + ct^2 \qquad\qquad 2.170$$

Here, $t = 0$ corresponds to $\vec{x}_2$, and $t = 1$ corresponds to $\vec{x}_1$. This leads to the equations

$$f(\vec{x}_2) = F(0) = a \qquad\qquad 2.171$$

$$f(\vec{x}_1) = F(1) = a + b + c \qquad\qquad 2.172$$

$$f(\vec{\sigma}) = F(r) = a + br + cr^2 \qquad\qquad 2.173$$

This leads to the matrix equation

$$\begin{pmatrix} 1 & 1 \\ r & r^2 \end{pmatrix}\begin{pmatrix} b \\ c \end{pmatrix} = \begin{pmatrix} f(\vec{x}_1) - f(\vec{x}_2) \\ f(\vec{\sigma}) - f(\vec{x}_2) \end{pmatrix} \qquad\qquad 2.174$$

Solving this,

$$\begin{pmatrix} b \\ c \end{pmatrix} = \frac{1}{r(r-1)}\begin{pmatrix} r^2 & -1 \\ -r & 1 \end{pmatrix}\begin{pmatrix} f(\vec{x}_1) - f(\vec{x}_2) \\ f(\vec{\sigma}) - f(\vec{x}_2) \end{pmatrix} \qquad\qquad 2.175$$

From this, the location of the minimum of the quadratic is

$$t = -\frac{b}{2c} \qquad\qquad 2.176$$

Quadratic Vector Interpolation

I-1   Dependent variable z

I-2   Independent variables $\vec{x}$

I-3   Objective function $f(\vec{x})$ relating $z = f(\vec{x})$

I-4 N initial states x_i

I-5   Method for choosing uniform random deviates on the range $[0,1)$

S-1   Iterate over each state $\vec{x}_i$

S-2   Compute $f(\vec{x}_i)$ for each state

Choose a random state $\vec{x}_j$ where $i \neq j$

S-3   Alternatively, choose three state vectors and compute

$$\vec{M} = \frac{1}{3}(\vec{x}_1 + \vec{x}_2 + \vec{x}_3)$$

S-4   Select a random $r \in [0,1)$

S-5   Compute $\vec{\sigma} = \vec{x}_j + r\vec{d}$

S-6   Compute $a = f(\vec{x}_j)$

S-7   Compute $b = \dfrac{\left[r^2\left(f(\vec{x}_i)-f(\vec{x}_j)\right)-\left(f(\vec{\sigma})-f(\vec{x}_j)\right)\right]}{r(r-1)}$

S-8   Compute $c = \dfrac{\left[-r\left(f(\vec{x}_i)-f(\vec{x}_j)\right)+\left(f(\vec{\sigma})-f(\vec{x}_j)\right)\right]}{r(r-1)}$

S-9   Compute $t = -\dfrac{b}{2c}$

S-10   Compute $\vec{\rho} = \vec{x}_j + t\vec{d}$

S-11   If $f(\vec{\rho})$ is better than $f(\vec{x}_i)$, replace $\vec{x}_i$ with $\vec{\rho}$ and process the next state

S-12   If $f(\vec{\sigma})$ is better than $f(\vec{x}_i)$, replace $\vec{x}_i$ with $\vec{\sigma}$ and process the next state

S-13   Repeat from S-3 as desired

O-1   The best value of $f(\vec{x}_i)$ is the potential optimum.

Algorithm XXX: Quadratic vector interpolation is useful for discovering an optimum for the objective function.

Vector interpolation can be extended to interpolation in higher dimensional spaces. In general, a n-dimensional problem allows interpolation up to a generalized n-dimensional quadratic.

Let the problem space be n-dimensional and perform interpolation on a m-dimensional space where $m \leq n$. The algorithm begins be selecting $m + 1$ state vectors from the population. Designate these vectors as $\vec{x}^i$ where $i = 0, 1, \dots m$. The superscript i indicates the ith vector, not the ith component of a vector. Components are represented as subscripts. Thus, x_j^i is the jth component of the ith vector.

Use the 0th vector as the base and compute the difference between each vector and this base vector. Set

$$\vec{\Delta}^i = \vec{x}^i - \vec{x}^0 \qquad 2.177$$

The m-dimensional space is defined in terms of the m-dimensional $\vec{t}$. The generalized quadratic in this space has the form

$$\psi(\vec{t}) = a + b^i t_i + c^{ij} t_i t_j \qquad 2.178$$

This above expression uses the Einstein summation convention where repeated indices indicate a sum. Thus, the second term represents the sum of m products (b is a vector), while the third term is m^2 terms. The third terms leads to expressions such as

$$(c^{12} + c^{21}) t_1 t_2 \qquad 2.179$$

There is no need for two coefficients multiplying the same expression. For simplicity, we assume $c^{ij} = 0$ whenever $j > i$. In this case, the third term results in a total of $\frac{m(m+1)}{2}$ remaining terms. In total the generalized quadratic has

$$1 + m + \frac{m(m+1)}{2} = \frac{(m+1)(m+2)}{2} \qquad 2.180$$

parameters.

The vector space defined by $\vec{t}$ may be used to specify vectors from the search space through a vector-valued map. In component form this is

$$F_k(\vec{t}) = x_k^0 + t_i \Delta_k^i \qquad 2.181$$

or

$$\vec{F}(\vec{t}) = \vec{x}^0 + \vec{t} \cdot \vec{\Delta}$$

2.182

In this expression, $\vec{\Delta}$ is the vector formed from the ith component of all of the difference vectors. Essentially, Δ is a matrix whose rows are the difference vectors. $\vec{\Delta}$ is the vector made from the columns of Δ.

Let ε_k be the kth unit vector in t-space. From the above expression,

$$\vec{F}(\vec{0}) = \vec{x}^0$$

2.183

$$\vec{F}(\varepsilon_k) = \vec{x}^k$$

2.184

In this sense the function F maps the search space to t-space. Using these values of t in the expression for the quadratic fit,

$$\psi(\vec{0}) = a = f(\vec{x}^0)$$

2.185

$$\psi(\varepsilon_k) = a + b^k + c^{kk} = f(\vec{x}^k)$$

2.186

This provides $m + 1$ equations relating the coefficients of the fit parameters. To compute the fit, we need at least

$$\frac{(m + 1)(m + 2)}{2} - (m + 1) = \frac{m(m + 1)}{2}$$

2.187

additional data points. These are chosen by the following process:

1. Generate vector in t-space whose components are independent uniform random deviates.
2. Compute the value of the vector in the search space using 2.182.
3. Compute the value of the objective function at the vector from (2).
4. Substitute the value of the objective function and the form of the random vector into 2.186 to provide another equation relating the fit parameters.
5. If this is the ith vector generated under this process, designate the random vector in t-space as $\vec{\rho}^i$, the corresponding vector in the search space as $\vec{r}^i$, and the value of the objective function as $f(\vec{r}^i)$.
6. Repeat from 1 for each of the $\frac{m(m+1)}{2}$ required points.

With this, we have enough equations to compute the fit parameters. First, note that the parameter a is trivially computed from 2.185:

$$a = f(\vec{x}^0)$$

2.188

Since this parameter is known already, we can eliminate it from the other equations. The resulting set of equation is then

$$m \qquad\qquad b^k + c^{kk} = f(\vec{x}^k) - f(\vec{x}^0) \qquad\qquad 2.189$$

$$\frac{m(m+1)}{2} \qquad\qquad b^i \rho_i^k + c^{ij} \rho_i^k \rho_j^k = f(\vec{r}^k) - f(\vec{x}^0) \qquad\qquad 2.190$$

In the above expressions, the value on the far left is the number of equations for each form. Here, the fit parameters b and c are the unknown variables. The above equation may be written in the form

$$C\vec{p} = \vec{d} \qquad\qquad 2.191$$

where

$$\vec{d} = \begin{pmatrix} f(\vec{x}^1) - f(\vec{x}^0) \\ f(\vec{x}^2) - f(\vec{x}^0) \\ \vdots \\ f(\vec{x}^m) - f(\vec{x}^0) \\ f(\vec{r}^1) - f(\vec{x}^0) \\ f(\vec{r}^2) - f(\vec{x}^0) \\ \vdots \\ f\left(\vec{r}^{\frac{m(m+1)}{2}}\right) - f(\vec{x}^0) \end{pmatrix} \qquad\qquad 2.192$$

$$\vec{p} = \begin{pmatrix} b^1 \\ b^2 \\ \vdots \\ b^m \\ c^{11} \\ c^{21} \\ c^{22} \\ c^{31} \\ \vdots \\ c^{mm} \end{pmatrix} \qquad\qquad 2.193$$

$$C = \begin{pmatrix} 1 & 0 & \cdots & 0 & 1 & 0 & \cdots & 0 \\ 0 & 1 & \cdots & 0 & 0 & 1 & \cdots & 0 \\ \vdots & \vdots & \vdots & \vdots & \vdots & \vdots & \vdots & \vdots \\ \rho_1^1 & \rho_2^1 & \cdots & \rho_m^1 & \rho_1^1\rho_1^1 & \rho_2^1\rho_1^1 & \cdots & \rho_m^1\rho_m^1 \\ \vdots & \vdots & \vdots & \vdots & \vdots & \vdots & \vdots & \vdots \\ \rho_1^m & \rho_2^m & \cdots & \rho_m^m & \rho_1^m\rho_1^m & \rho_2^m\rho_1^m & \cdots & \rho_m^m\rho_m^m \end{pmatrix} \qquad\qquad 2.194$$

This may be solved by inverting the matrix C. Formally,

$$\vec{p} = C^{-1}\vec{d} \qquad\qquad 2.195$$

This provides the values for the fit parameters. From this, we can compute the optimum for the fit by taking the partial derivatives with respect to the t-variables in 2.178 and setting them to zero:

$$\frac{\partial \psi(\vec{t})}{\partial t_k} = b^k + \left(c^{kj} + \delta_k^j\right) t_j = 0 \qquad \text{2.196}$$

In this expression, δ_k^j is the Kronecker delta function. This term arises from the partial derivative of the term $(t_k)^2$. The above expression is a system of m equations relating the m unknown t-varaibles. This system may be written as

$$\mathcal{A}\vec{t}^* = -\vec{b} \qquad \text{2.197}$$

where $\vec{t}^*$ is the point in t-space for the optimum value of the quadratic fit, $\vec{b}$ is the vector formed from the first m elements of $\vec{p}$, and the matrix $\mathcal{A}$ is

$$\mathcal{A} = \begin{pmatrix} p_0 p_0 & p_0 p_1 & \cdots & p_0 p_m \\ p_1 p_0 & p_1 p_1 & \cdots & p_1 p_m \\ \vdots & \vdots & \vdots & \vdots \\ p_m p_0 & p_m p_1 & \cdots & p_m p_m \end{pmatrix} \qquad \text{2.198}$$

The formal solution is

$$\vec{t}^* = -\mathcal{A}^{-1}\vec{b} \qquad \text{2.199}$$

From this we have computed the location in t-space of the optimum for the fit function. Using 2.182, the corresponding point in the search space is

$$\vec{F}(\vec{t}) = \vec{x}^0 + \vec{t}^* \cdot \vec{\Delta} = \vec{x}^* \qquad \text{2.200}$$

Next, compute the value of the objective function at this point $f(\vec{x}^*)$. Based on this, the population is updated as follows:

1. Find the vector $\vec{x}^k$ with the worst value $f(\vec{x}^k)$.
2. If $f(\vec{x}^*)$ is better than $f(\vec{x}^k)$, replace $\vec{x}^k$ with $\vec{x}^*$.
3. Find the random vector $\vec{r}^l$ with the best value $f(\vec{r}^l)$.
4. If $f(\vec{r}^l)$ is better than $f(\vec{x}^k)$, replace $\vec{x}^k$ with $\vec{r}^l$.
5. If $f(\vec{x}^*)$ and all of the $f(\vec{r}^l)$ are worse than $f(\vec{x}^k)$, then
   a. Find the vector $\vec{x}^b$ with the best value $f(\vec{x}^b)$.
   b. Replace all $\vec{x}^i$ except $\vec{x}^b$ using the formula

$$\vec{x}^i = \vec{x}^b + \sigma\left(\vec{x}^i - \vec{x}^b\right) \qquad \text{2.201}$$

m-Dimensional Vector Interpolation

I-1 Dependent variable z

I-2 Independent variables $\vec{x}$

I-3 Objective function $f(\vec{x})$ relating $z = f(\vec{x})$

I-4 N initial states x_i

I-5 Value of the fit space m where m is less than or equal to the dimension of the search space

I-6 Specify the value σ

S-1 Select $m + 1$ vectors from the population $\vec{x}^i$

S-2 Compute $f(\vec{x}^i)$ for each state

S-3 Compute the difference vectors
$$\vec{\Delta}^i = \vec{x}^i - \vec{x}^0$$

S-4 Compute $\vec{\rho}$ as a m-dimensional vector whose components are uniform random deviates

S-5 Compute the vector $\vec{r}^i = \vec{x}^0 + \vec{\rho} \cdot \vec{\Delta}$

S-6 Compute the value of the objective function $f(\vec{r}^i)$

S-7 Repeat from S-4 to generate $\dfrac{m(m+1)}{2}$ vectors

S-8 Compute

$$\vec{d} = \begin{pmatrix} f(\vec{x}^1) - f(\vec{x}^0) \\ f(\vec{x}^2) - f(\vec{x}^0) \\ \vdots \\ f(\vec{x}^m) - f(\vec{x}^0) \\ f(\vec{r}^1) - f(\vec{x}^0) \\ f(\vec{r}^2) - f(\vec{x}^0) \\ \vdots \\ f\left(\vec{r}^{\frac{m(m+1)}{2}}\right) - f(\vec{x}^0) \end{pmatrix}$$

S-9 Compute

$$C = \begin{pmatrix} 1 & 0 & \cdots & 0 & 1 & 0 & \cdots & 0 \\ 0 & 1 & \cdots & 0 & 0 & 1 & \cdots & 0 \\ \vdots & \vdots & \vdots & \vdots & \vdots & \vdots & \vdots & \vdots \\ \rho_1^1 & \rho_2^1 & \cdots & \rho_m^1 & \rho_1^1\rho_1^1 & \rho_2^1\rho_1^1 & \cdots & \rho_m^1\rho_m^1 \\ \vdots & \vdots & \vdots & \vdots & \vdots & \vdots & \vdots & \vdots \\ \rho_1^m & \rho_2^m & \cdots & \rho_m^m & \rho_1^m\rho_1^m & \rho_2^m\rho_1^m & \cdots & \rho_m^m\rho_m^m \end{pmatrix}$$

S-10   Compute $\vec{p} = C^{-1}\vec{d}$

S-11   Compute $\vec{b}$ as the first m components of $\vec{p}$

S-12   Compute

$$\mathcal{A} = \begin{pmatrix} p_0p_0 & p_0p_1 & \cdots & p_0p_m \\ p_1p_0 & p_1p_1 & \cdots & p_1p_m \\ \vdots & \vdots & \vdots & \vdots \\ p_mp_0 & p_mp_1 & \cdots & p_mp_m \end{pmatrix}$$

S-13   Compute

$$\vec{t}^* = -\mathcal{A}^{-1}\vec{b}$$

S-14   Compute

$$\vec{x}^* = \vec{x}^0 + \vec{t}^* \cdot \vec{\Delta}$$

S-15   Compute $f(\vec{x}^*)$

S-16   Identify the vector in the set $\vec{x}^*$, $\vec{r}^l$ that has the best value for the objective function. Designate this vector as $\vec{x}'$

S-17   Identify the vector in the set $\vec{x}^k$ that has the worst value for the objective function. Designate this vector as $\vec{x}^w$

S-18   If $f(\vec{x}')$ is better than $f(\vec{x}^w)$, replace $\vec{x}^w$ with $\vec{x}'$, repeat from S-1.

S-19   Identify the vector in the set $\vec{x}^k$ that has the best value for the objective function. Designate this vector as $\vec{x}^b$

S-20   Replace all vectors $\vec{x}^k$ except $\vec{x}^b$ using the formula

$$\vec{x}^k = \vec{x}^b + \sigma(\vec{x}^k - \vec{x}^b)$$

S-21   Repeat from S-1

O-1   The best value of $f(\vec{x}_i)$ is the potential optimum.

Algorithm XXXI: m-dimensional vector interpolation is useful for discovering an optimum for the objective function.

3 Performance Measures

3.1 Measuring Performance of Optimization Algorithms

In the last chapter we identified several methods for computing the global optimum for an objective function. Each of these methods has strengths and weaknesses, and each is useful with certain problems and less useful on others.

In this chapter we will apply these techniques against a standard set of objective functions and measure the performance of each technique. The basic measure of performance used is the number of times the objective function must be evaluated to arrive at the desired level of accuracy.

Several objective functions are listed in the tables below. Each objective function is specified as a function over a vector $\vec{x}$ along with the search space under consideration. For each objective function we also specify the global minimum and a target window. When a method finds at least one test point in the target window, this is considered as successfully identifying the global minimum. In addition, many of these functions have many local optima that can trap optimization algorithms far from the global optima.

3.2 Objective Functions

ACKLEY FUNCTION			
Definition	$f(\vec{x}) = 20 + e - 20 \exp\left(-\frac{1}{5}\sqrt{\frac{1}{n}\vec{x} \cdot \vec{x}}\right) - \exp\left(\frac{1}{n}\sum_{i=1}^{n} \cos(2\pi x_i)\right)$		

	Search Domain	$-15 \leq x_i \leq 30$	
	Global Optima	$\vec{o} = \vec{0}$	
		Value	0
	Tolerance	$|\vec{x}| < 10^{-6}$	**Dimension**
			n

ALUFFI-PENTINI FUNCTION			
Definition	$f(x,y) = \dfrac{1}{4}x^4 + \dfrac{1}{2}(y^2 - x^2) + \dfrac{1}{10}x$		
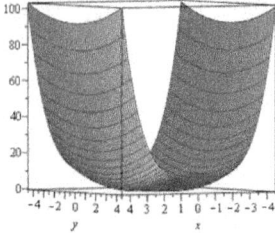	**Search Domain**	$-10 \le x_i \le 10$	
	Global Optima	$\vec{o} = (-1.0465, 0)$	
		Value	-0.3523
	Tolerance	$|\vec{x} - \vec{o}| < 10^{-8}$	Dimension
			2

BEALE FUNCTION			
Definition	$f(x,y) = (1.5 - x + xy)^2 + (2.25 - x + xy^2)^2 + (2.625 - x + xy^3)^2$		
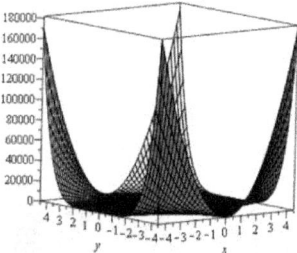	**Search Domain**	$-4.5 \le x_i \le 4.5$	
	Global Optima	$\vec{o} = (3, 0.5)$	
		Value	0
	Tolerance	$|\vec{x} - \vec{o}| < 10^{-8}$	Dimension
			2

BECKER AND LAGO FUNCTION			
Definition	$f(x,y) = (|x| - 5)^2 + (|y| - 5)^2 + .1[(x - 5)^2 + (y - 5)^2]$		
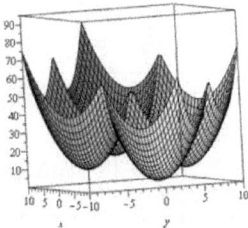	**Search Domain**	$-10 \le x_i \le 10$	
	Global Optima	$\vec{o} = (5, 5)$	
		Value	0
	Tolerance	$|\vec{x} - \vec{o}| < 10^{-8}$	Dimension
			2

BIBESSEL FUNCTION			
Definition	$f(x,y) = -J_0(x)J_0(y)$		

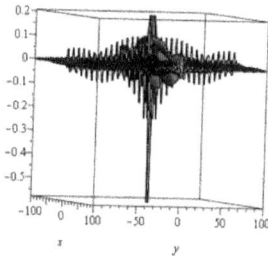

	Search Domain	$-100 \leq x_i \leq 100$	
	Global Optima	$\vec{o} = (0,0)$	
		Value	0
	Tolerance	$\|\vec{x} - \vec{o}\| < 10^{-8}$	Dimension
			2

BOHACHEVSKY I FUNCTION			
$f(x,y) = x^2 + 2y^2 - .3\cos(3\pi x) - .4\cos(4\pi y) + .7$			

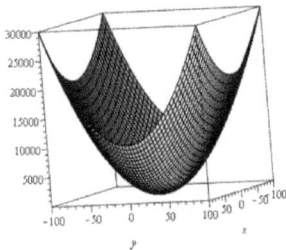

	Search Domain	$-100 \leq x_i \leq 100$	
	Global Optima	$\vec{o} = \vec{0}$	
		Value	0
	Tolerance	$\|\vec{x} - \vec{o}\| < 10^{-8}$	Dimension
			2

BOHACHEVSKY II FUNCTION			
$f(x,y) = x^2 + 2y^2 - .3\cos(3\pi x)\cos(4\pi y) + .3$			

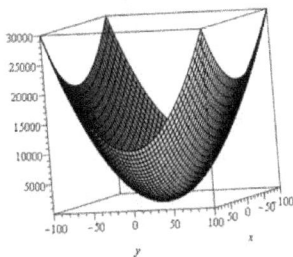

	Search Domain	$-100 \leq x_i \leq 100$	
	Global Optima	$\vec{o} = \vec{0}$	
		Value	0
	Tolerance	$\|\vec{x} - \vec{o}\| < 10^{-6}$	Dimension
			2

BOHACHEVSKY III FUNCTION

$$f(x,y) = x^2 + 2y^2 - .3\cos(3\pi x + 4\pi y) + .3$$

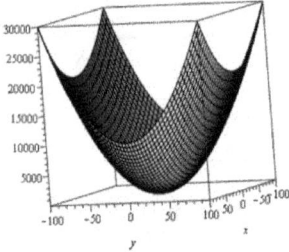

Search Domain	$-100 \le x_i \le 100$	
Global Optima	$\vec{o} = \vec{0}$	
	Value	0
Tolerance	$\lvert \vec{x} - \vec{o} \rvert < 10^{-8}$	Dimension
		2

BOOTH FUNCTION

Definition	$f(x,y) = (x + 2y - 7)^2 + (2x + y - 5)^2$

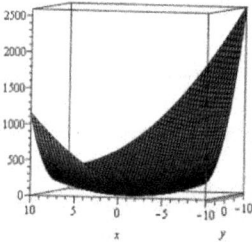

Search Domain	$-10 \le x_i \le 10$	
Global Optima	$\vec{o} = (1,3)$	
	Value	0
Tolerance	$\lvert \vec{x} - \vec{o} \rvert < 10^{-8}$	Dimension
		2

BRANIN FUNCTION

Definition	$f(x,y) = (y - 1.275\pi^{-2}x^2 + 5\pi^{-1}x - 6)^2 + 10(1 - (8\pi)^{-1})\cos x + 10$ $+ .1[(x - \pi)^2 + (y - 2.275)^2]$

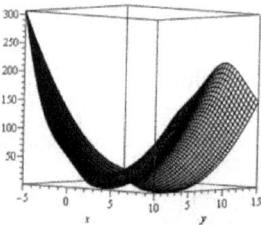

Search Domain	$-10 \le x_i \le 10$	
Global Optima	$\vec{o} = (\pi, 2.275)$	
	Value	0.397887
Tolerance	$\lvert \vec{x} - \vec{o} \rvert < 10^{-8}$	Dimension
		2

COLVILLE FUNCTION

Definition	$f(x,y,z,w) = 100(x^2 - y^2)^2 + (x-1)^2 + (z-1)^2 + 90(z^2 - w)^2$ $+ 10.1[(y-1)^2 + (w-1)^2] + 19.8(y-1)(w-1)$		

	Search Domain	$-10 \le x_i \le 10$	
	Global Optima	$\vec{o} = \vec{1}$	
		Value	0
	Tolerance	$|\vec{x} - \vec{o}| < 10^{-6}$	Dimension
			4

COSINE MIX FUNCTION

Definition	$f(\vec{x}) = -.1 \sum_{i=1}^{n} \cos(5\pi x_i) + \sum_{i=1}^{n} x_i^2$		

	Search Domain	$-1 \le x_i \le 1$	
	Global Optima	$\vec{o} = \vec{0}$ $n = 2$	
		Value	0.2
	Tolerance	$|\vec{x} - \vec{o}| < 10^{-6}$	Dimension
			n

DEKKERS AND ARTS FUNCTION

Definition	$f(x,y) = 10^5 x^2 + y^2 - (x^2 + y^2)^2 + 10^{-5}(x^2 + y^2)^4 + .1[x^2 + (y-15)^2]$		

	Search Domain	$-20 \le x_i \le 20$	
	Global Optima	$\vec{o} = (0, 14.94511523)$	
		Value	-24776.51804
	Tolerance	$|\vec{x} - \vec{o}| < 10^{-8}$	Dimension
			2

DIXON & PRICE FUNCTION			
Definition	$f(\vec{x}) = (x_1 - 1)^2 + \displaystyle\sum_{i=2}^{n} i(2x_i^2 - x_{i-1})^2$		
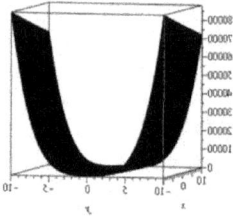	**Search Domain**	$-10 \le x_i \le 10$	
	Global Optima	$\vec{o} = \vec{0}$	
		Value	0
	Tolerance	$|\vec{x} - \vec{o}| < 10^{-8}$	Dimension
			n

EASOM FUNCTION			
Definition	$f(x,y) = -\cos(x)\cos(y)\exp(-(x-\pi)^2 - (y-\pi)^2)$		
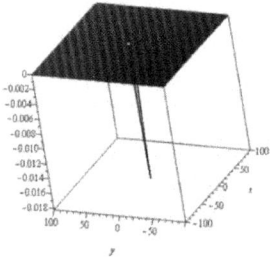	**Search Domain**	$-100 \le x_i \le 100$	
	Global Optima	$\vec{o} = (\pi, \pi)$	
		Value	-1
	Tolerance	$|\vec{x} - \vec{o}| < 10^{-8}$	Dimension
			2

EXPONENTIAL FUNCTION			
Definition	$f(x,y) = 1 - \exp\left(-\dfrac{1}{2}\displaystyle\sum_{i=1}^{n} x_i^2\right)$		
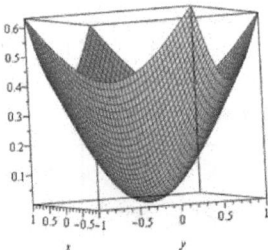	**Search Domain**	$-1 \le x_i \le 1$	
	Global Optima	$\vec{o} = \vec{0}$	
		Value	0
	Tolerance	$|\vec{x} - \vec{o}| < 10^{-8}$	Dimension
			n

GOLDSTEIN & PRICE FUNCTION

$$f(x,y,z,w) = [1 + (x+y+1)^2(19 - 14x + 3x^2 - 14y + 6xy + 3y^2)][30 + (2x-3y)^2(18 - 32x + 12x^2 + 48y - 36xy + 27y^2)]$$

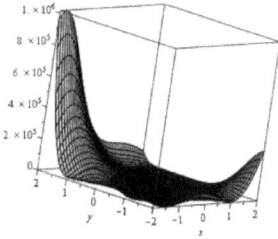

Search Domain	$-2 \leq x_i \leq 2$	
Global Optima	$\vec{o} = (0,-1)$	
	Value	3
Tolerance	$\|\vec{x} - \vec{o}\| < 10^{-8}$	Dimension
		2

GRIEWANK FUNCTION

$$f(\vec{x}) = 1 + \sum_{i=1}^{n} \frac{x_i^2}{4000} - \prod_{i=1}^{n} \cos\left(\frac{x_i}{\sqrt{i}}\right)$$

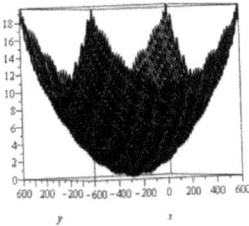

Search Domain	$-600 \leq x_i \leq 600$	
Global Optima	$\vec{o} = \vec{0}$	
	Value	0
Tolerance	$\|\vec{x} - \vec{o}\| < 10^{-8}$	Dimension
		n

HELICAL VALLEY FUNCTION

$$f(x,y,z) = 100\left[(y - 10\theta)^2 + \left(\sqrt{x^2 + y^2} - 1\right)^2\right] + z^3$$

$$\theta = \frac{1}{2\pi}\arctan\frac{y}{x} + \frac{1}{2}H(-x) \quad H(x) \text{ is the Heaviside step function}$$

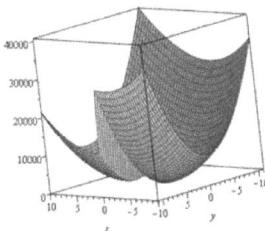

Search Domain	$-10 \leq x_i \leq 10$	
Global Optima	$\vec{o} = (1,0,0)$	
	Value	0
Tolerance	$\|\vec{x} - \vec{o}\| < 10^{-8}$	Dimension
		3

HOSAKI FUNCTION		
$f(x,y) = \left(1 - 8x + 7x^2 - \dfrac{7}{3}x^3 + \dfrac{1}{4}x^4\right)y^2 e^{-y}$		

<table>
<tr><td rowspan="6">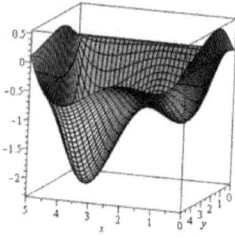</td><td>Search Domain</td><td colspan="2" align="center">$0 \le x_i \le 5$</td></tr>
<tr><td rowspan="2">Global Optima</td><td colspan="2" align="center">$\vec{o} = (4,2)$</td></tr>
<tr><td>Value</td><td align="center">-2.3458</td></tr>
<tr><td rowspan="3">Tolerance</td><td rowspan="3">$|\vec{x} - \vec{o}| < 10^{-8}$</td><td align="center">Dimension</td></tr>
<tr><td></td></tr>
<tr><td align="center">2</td></tr>
</table>

HIMMELBLAU FUNCTION		
$f(x,y) = (x + y^2 - 7)^2 + (x^2 + y - 11)^2 + .1[(x - 3)^2 + (y - 2)^2]$		

<table>
<tr><td rowspan="6">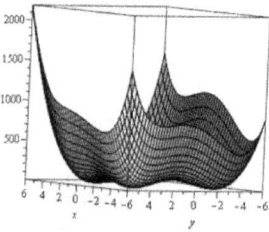</td><td>Search Domain</td><td colspan="2" align="center">$-6 \le x_i \le 6$</td></tr>
<tr><td rowspan="2">Global Optima</td><td colspan="2" align="center">$\vec{o} = (3,2)$</td></tr>
<tr><td>Value</td><td align="center">0</td></tr>
<tr><td rowspan="3">Tolerance</td><td rowspan="3">$|\vec{x} - \vec{o}| < 10^{-8}$</td><td align="center">Dimension</td></tr>
<tr><td></td></tr>
<tr><td align="center">2</td></tr>
</table>

HUMP I FUNCTION		
$f(x,y) = 4x^2 - 2.1x^4 + \dfrac{1}{3}x^6 + xy - 4y^2 + 4y^4 + .1[(x - .089842)^2 + (y + .712656)^2]$		

<table>
<tr><td rowspan="6">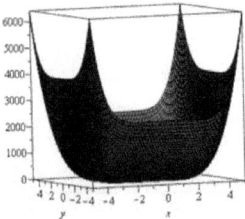</td><td>Search Domain</td><td colspan="2" align="center">$-5 \le x_i \le 5$</td></tr>
<tr><td rowspan="2">Global Optima</td><td colspan="2" align="center">$\vec{o} = (0.089842, -0.712656)$</td></tr>
<tr><td>Value</td><td align="center">-1.0316</td></tr>
<tr><td rowspan="3">Tolerance</td><td rowspan="3">$|\vec{x} - \vec{o}| < 10^{-8}$</td><td align="center">Dimension</td></tr>
<tr><td></td></tr>
<tr><td align="center">2</td></tr>
</table>

HUMP II FUNCTION

Definition	$f(x,y) = 2x^2 - 1.05x^4 + \dfrac{1}{3}x^6 + xy + y^2$		

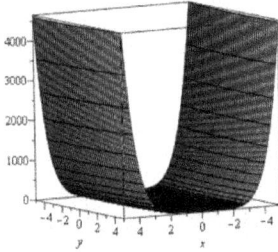

	Search Domain	$-5 \le x_i \le 5$	
	Global Optima	$\vec{o} = (0,0)$	
		Value	0
	Tolerance	$\|\vec{x} - \vec{o}\| < 10^{-8}$	Dimension
			2

LEVY FUNCTION

$$f(x,y,z,w) = \sin^2(\pi x_1) + (x_n - 1)^2(1 + 10\sin^2(2\pi x_n)) + \sum_{i=1}^{n-1}(x_i - 1)^2(1 + 10\sin^2(1 + \pi x_i))$$

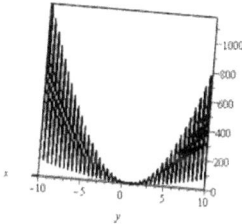

	Search Domain	$-10 \le x_i \le 10$	
	Global Optima	$\vec{o} = (1,1 \ldots,1)$	
		Value	0
	Tolerance	$\|\vec{x} - \vec{o}\| < 10^{-6}$	Dimension
			n

MATYAS FUNCTION

$$f(\vec{x}) = .26(x^2 + y^2) - .48xy$$

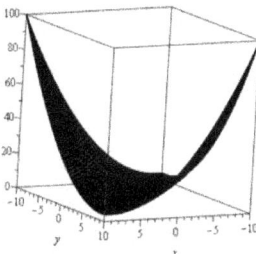

	Search Domain	$-10 \le x_i \le 10$	
	Global Optima	$\vec{o} = \vec{0}$	
		Value	0
	Tolerance	$\|\vec{x} - \vec{o}\| < 10^{-8}$	Dimension
			2

McCormick Function

$$f(\vec{x}) = \sin(x + y) + (x - y)^2 - \frac{1}{2}(3x - 5y) + 1$$

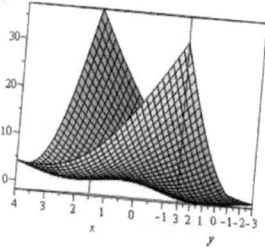

	Search Domain	$-1.5 \leq x \leq 4$ $-3 \leq x \leq 3$	
	Global Optima	$\vec{o} = (-0.574, -1.547)$	
		Value	-1.91219
	Tolerance	$|\vec{x} - \vec{o}| < 10^{-8}$	Dimension
			2

Michalewicz Function

$$f(x, y) = -\sum_{i=1}^{n} \sin(x_i) \sin^{20}\left(\frac{ix_i^2}{\pi}\right)$$

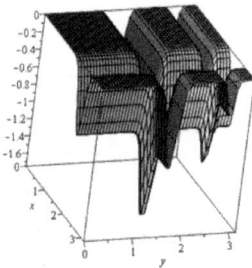

	Search Domain	$0 \leq x_i \leq \pi$	
	Global Optima	Varies with n $\vec{o} = (2.2029055, 1)$ $n = 2$	
		Value	-0.801329
	Tolerance	$|\vec{x} - \vec{o}| < 10^{-8}$	Dimension
			n

Miele and Cantrell Function

$$f(x, y, z, w) = (e^x - y)^4 + 100(y - z)^6 + \tan^4(z - w) + x^8$$

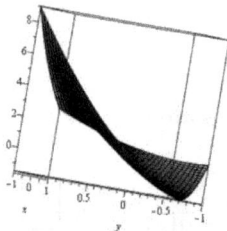

	Search Domain	$-1 \leq x_i \leq 1$	
	Global Optima	$\vec{o} = (0, 1, 1, 1)$	
		Value	0
	Tolerance	$|\vec{x} - \vec{o}| < 10^{-8}$	Dimension
			4

NEUMAIER FUNCTION

$$f(x,y,z,w) = \left(8 - \sum_{i=1}^{4} x_i\right)^2 + \left(18 - \sum_{i=1}^{4} x_i^2\right)^2 + \left(44 - \sum_{i=1}^{4} x_i^3\right)^2 + \left(114 - \sum_{i=1}^{4} x_i^4\right)^2$$

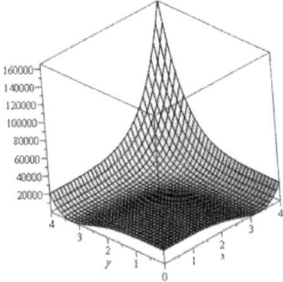

Search Domain	$0 \le x_i \le 4$
Global Optima	$\vec{o} = (1,2,2,3)$
Value	0
Tolerance	$\|\vec{x} - \vec{o}\| < 10^{-8}$ — Dimension: 4

PAVIANI FUNCTION

$$f(\vec{x}) = \sum_{i=1}^{n} [\ln^2(x_i - 2) + \ln^2(10 - x_i)] - \left[\prod_{i=1}^{n} x_i\right]^{.2}$$

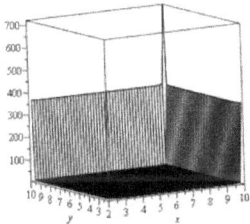

Search Domain	$2 \le x_i \le 10$
Global Optima	$\vec{o} = (\alpha, \alpha, \dots, \alpha)$ $\alpha = 9.3502$ $n = 10$
Value	-45.778
Tolerance	$\|\vec{x} - \vec{o}\| < 10^{-8}$ — Dimension: n

PERIODIC FUNCTION

$$f(x,y) = 1 + \sin^2 x + \sin^2 y - .1 \exp(-x^2 - y^2)$$

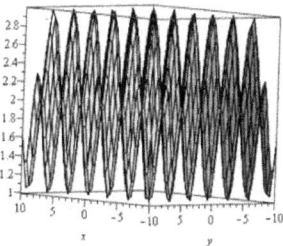

Search Domain	$-10 \le x_i \le 20$
Global Optima	$\vec{o} = \vec{0}$
Value	0.9
Tolerance	$\|\vec{x} - \vec{o}\| < 10^{-6}$ — Dimension: 2

PERM FUNCTION

$$f(x,y,z,w) = \sum_{k=1}^{n}\left[\sum_{i=1}^{n}(i^k + .5)\left(\left(\frac{x_i}{i}\right)^k - 1\right)\right]^2$$

Search Domain	$-n \le x_i \le n$		
Global Optima	$\vec{o} = (1,2 \dots, n)$		
	Value	0	
Tolerance	$|\vec{x} - \vec{o}| < 10^{-6}$	Dimension	
		n	

RASTRIGIN FUNCTION

$$f(\vec{x}) = 10n + \sum_{i=1}^{n}\left(x_i^2 - 10\cos(2\pi x_i)\right)$$

Search Domain	$-5.12 \le x_i \le 5.12$		
Global Optima	$\vec{o} = \vec{0}$		
	Value	0	
Tolerance	$|\vec{x} - \vec{o}| < 10^{-8}$	Dimension	
		n	

ROSENBROCK FUNCTION

$$f(x,y) = \sum_{i=1}^{n-1}\left[100\left(x_i^2 - x_{i+1}\right)^2 + (x_i - 1)^2\right]$$

Search Domain	$-5 \le x_i \le 10$		
Global Optima	$\vec{o} = (1,1, \dots, 1)$		
	Value	0	
Tolerance	$|\vec{x} - \vec{o}| < 10^{-8}$	Dimension	
		n	

SALOMON FUNCTION

$$f(\vec{x}) = -\cos\left(2\pi\sqrt{\sum_{i=1}^{n}x_i^2}\right) + 1 + .1\sqrt{\sum_{i=1}^{n}x_i^2}$$

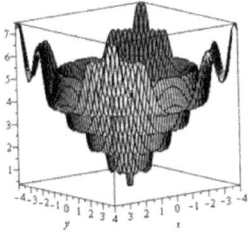

Search Domain	$-4 \le x_i \le 4$	
Global Optima	$\vec{o} = \vec{0}$	
	Value	0
Tolerance	$|\vec{x} - \vec{o}| < 10^{-8}$	Dimension
		n

SCHWEFEL FUNCTION

$$f(\vec{x}) = 418.9828872n - \sum_{i=1}^{n}x_i \sin\sqrt{|x_i|}$$

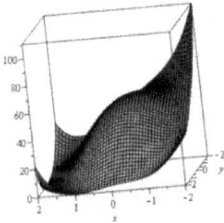

Search Domain	$-500 \le x_i \le 500$	
Global Optima	$\vec{o} = (\alpha, \alpha, \dots, \alpha)$ $\alpha = 420.9687464$	
	Value	0
Tolerance	$|\vec{x} - \vec{o}| < 10^{-6}$	Dimension
		n

SHEKEL'S FUNCTION

$$f(\vec{x}) = -\sum_{j=1}^{10}\left[c_j + \sum_{i=1}^{4}(x_j - a_{ij})^2\right]^{-1} \quad c_j = (.1,.2,.2,.4,.4,.6,.3,.7,.5,.5) \quad a^T = \begin{pmatrix} 4 & 1 & 8 & 6 & 3 & 2 & 5 & 8 & 6 & 7 \\ 4 & 1 & 8 & 6 & 7 & 9 & 5 & 1 & 2 & 3.6 \\ 4 & 1 & 8 & 6 & 3 & 2 & 3 & 8 & 6 & 7 \\ 4 & 1 & 8 & 6 & 7 & 9 & 3 & 1 & 2 & 3.6 \end{pmatrix}$$

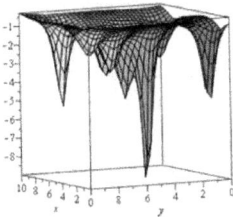

Search Domain	$0 \le x_i \le 10$	
Global Optima	$\vec{o} = (4,4,\dots,4)$	
	Value	-10.53628372622
Tolerance	$|\vec{x} - \vec{o}| < 10^{-6}$	Dimension
		4

SPHERE FUNCTION

$$f(\vec{x}) = \sum_{i=1}^{n} x_i^2$$

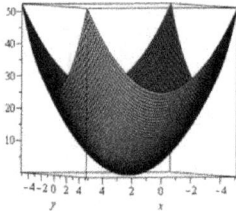

	Search Domain	$-5.12 \leq x_i \leq 5.12$	
	Global Optima	$\vec{o} = \vec{0}$	
		Value	0
	Tolerance	$|\vec{x} - \vec{o}| < 10^{-8}$	Dimension
			n

STEP FUNCTION

$$f(\vec{x}) = 6n + \sum_{i=1}^{n} \lfloor x_i \rfloor$$

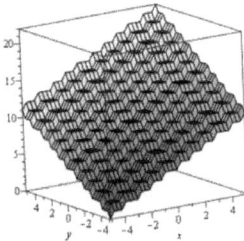

	Search Domain	$-5.12 \leq x_i \leq 5.12$	
	Global Optima	$\vec{o} = \vec{0}$	
		Value	0
	Tolerance	$|\vec{x} - \vec{o}| < 10^{-8}$	Dimension
			n

SUM SQUARES FUNCTION

$$f(x, y) = \sum_{i=1}^{n} i x_i^2$$

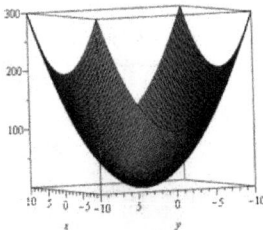

	Search Domain	$-10 \leq x_i \leq 10$	
	Global Optima	$\vec{o} = \vec{0}$	
		Value	0
	Tolerance	$|\vec{x} - \vec{o}| < 10^{-8}$	Dimension
			n

TRID FUNCTION

$$f(x,y,z,w) = \sum_{i=1}^{n}(x_i - 1)^2 - \sum_{i=2}^{n}x_i x_{i-1}$$

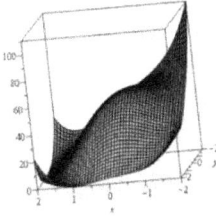

Search Domain	$-n^2 \le x_i \le n^2$		
Global Optima	$\vec{o} = (2,2) \quad n = 2$		
	Value		-2
Tolerance	$|\vec{x} - \vec{o}| < 10^{-6}$	**Dimension**	
		n	

WHITLEY FUNCTION

$$f(x,y) = \sum_{i=1}^{n}\sum_{j=1}^{n}\left[\frac{\left(100\left(x_i^2 - x_j\right)^2 + \left(1 - x_j\right)^2\right)^2}{4000} - \cos\left(100\left(x_i^2 - x_j\right)^2 + \left(1 - x_j\right)^2\right) + 1\right]$$

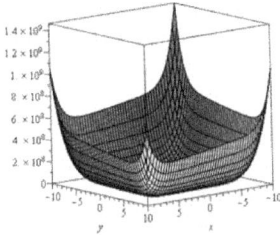

Search Domain	$-10 \le x_i \le 10$		
Global Optima	$\vec{o} = \vec{1}$		
	Value		0
Tolerance	$|\vec{x} - \vec{o}| < 10^{-6}$	**Dimension**	
		n	

ZAKHAROV FUNCTION

$$f(\vec{x}) = \sum_{i=1}^{n}x_i^2 + \left(\sum_{i=1}^{n}.5ix_i\right)^2 + \left(\sum_{i=1}^{n}.5ix_i\right)^4$$

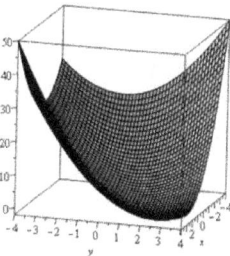

Search Domain	$-5 \le x_i \le 10$		
Global Optima	$\vec{o} = \vec{0}$		
	Value		0
Tolerance	$|\vec{x} - \vec{o}| < 10^{-8}$	**Dimension**	
		n	

3.3 Testing Optimization Algorithms

For measuring performance in this chapter, we examine algorithms that are applicable to multivariate objective functions, and do not require computing a derivative. These requirements provide a wide range of applicability for the objective function. The optimization algorithms that meet these criteria are shown in the table below.

Optimization Methods		
Particle Swam Optimization (PSO)	Genetic Algorithm (GA)	Differential Evolution (DE)
Quantum Particle Search (QPS)	Amoeba Method (AM)	Simulated Annealing (SA)

We test each of these methods using the test functions from the previous section. Most of these objective functions were originally designed to test deterministic optimization methods. However, all of the above methods are stochastic except the relaxation and the amoeba methods. Each of the methods use one or more random starting points.

Because of the stochastic nature of these algorithms, it is important to control for situations where a method quickly converges to the optimum simply because the initial set of random vectors was particularly good. Therefore, rather than simply computing the number of function evaluations on a single trial, we compute multiple trials for each method. From these results we examine the mean, variance, and the statistical distribution of the number of objective function evaluations.

The method for test proceeds as follows. Each algorithm is applied to each objective function 10,000 times. The algorithm is applied until at least one test point is within the radius of convergence. For each algorithm-function combination, the average and variance of the number of function evaluations required for convergence is computed.

If the algorithm fails to converge after 10 million evaluations, the attempt is thrown out. Thus, the average and variance are computed only for the attempts that converge. We record the number of times the algorithm converged successfully over the 10,000 trials.

Algorithm	Problem	Avg	Conv.	Algorithm	Problem	Avg	Conv.
AM	Ackley	1000	1.00	AM	HumpII	442	1.00
DE	Ackley	10309	1.00	DE	HumpII	2121	1.00
GA	Ackley	70951	0.87	GA	HumpII	1880	1.00
SA	Ackley	NA	0.00	SA	HumpII	799	1.00
PS	Ackley	2816	0.04	PS	HumpII	396	1.00
QPS	Ackley	33486	1.00	QPS	HumpII	532	1.00
VI	Ackley	NA	0.00	VI	HumpII	109	1.00
VII	Ackley	NA	0.00	VII	HumpII	347	1.00
VInD	Ackley	NA	0.00	VInD	HumpII	134	1.00
AM	Aluffi	399	1.00	AM	Levy	1229	0.08
DE	Aluffi	2604	1.00	DE	Levy	8541	1.00
GA	Aluffi	3542	1.00	GA	Levy	74133	0.71
SA	Aluffi	1107	1.00	SA	Levy	NA	0.00
PS	Aluffi	576	1.00	PS	Levy	NA	0.00
QPS	Aluffi	723	1.00	QPS	Levy	21714	1.00
VI	Aluffi	224	1.00	VI	Levy	NA	0.00
VII	Aluffi	813	1.00	VII	Levy	NA	0.00
VInD	Aluffi	294	1.00	VInD	Levy	NA	0.00
AM	Beale	357	0.93	AM	Matyas	508	1.00
DE	Beale	5223	1.00	DE	Matyas	13010	1.00
GA	Beale	2831	1.00	GA	Matyas	11142	1.00
SA	Beale	1062	1.00	SA	Matyas	1357	1.00
PS	Beale	508	1.00	PS	Matyas	579	1.00
QPS	Beale	1414	1.00	QPS	Matyas	5742	1.00
VI	Beale	267	1.00	VI	Matyas	93	1.00
VII	Beale	1024	1.00	VII	Matyas	612	1.00
VInD	Beale	409	1.00	VInD	Matyas	54	1.00
AM	Becker	492	1.00	AM	McCormick	80	0.11
DE	Becker	2813	1.00	DE	McCormick	2800	1.00
GA	Becker	2446	1.00	GA	McCormick	903	1.00
SA	Becker	1476	1.00	SA	McCormick	747	1.00
PS	Becker	562	1.00	PS	McCormick	320	1.00
QPS	Becker	762	1.00	QPS	McCormick	411	1.00
VI	Becker	229	1.00	VI	McCormick	117	1.00
VII	Becker	1411	1.00	VII	McCormick	422	1.00
VInD	Becker	1000	1.00	VInD	McCormick	133	0.99

Algorithm	Problem	Avg	Conv.	Algorithm	Problem	Avg	Conv.
AM	BohachevskyI	840	1.00	AM	Michalewicz	105	0.84
DE	BohachevskyI	5508	1.00	DE	Michalewicz	880	0.88
GA	BohachevskyI	25534	1.00	GA	Michalewicz	307	1.00
SA	BohachevskyI	3410	1.00	SA	Michalewicz	618	0.82
PS	BohachevskyI	1202	1.00	PS	Michalewicz	538	0.95
QPS	BohachevskyI	3457	1.00	QPS	Michalewicz	300	0.93
VI	BohachevskyI	153	1.00	VI	Michalewicz	145	0.85
VII	BohachevskyI	1449	1.00	VII	Michalewicz	235	0.86
VInD	BohachevskyI	57	1.00	VInD	Michalewicz	159	0.83
AM	BohachevskyII	880	1.00	AM	Miele	683	1.00
DE	BohachevskyII	5599	1.00	DE	Miele	20674	0.32
GA	BohachevskyII	23248	0.53	GA	Miele	33309	0.26
SA	BohachevskyII	3397	1.00	SA	Miele	419	1.00
PS	BohachevskyII	1192	1.00	PS	Miele	558	1.00
QPS	BohachevskyII	4378	1.00	QPS	Miele	1019	1.00
VI	BohachevskyII	148	1.00	VI	Miele	1751	1.00
VII	BohachevskyII	1463	1.00	VII	Miele	793	1.00
VInD	BohachevskyII	56	1.00	VInD	Miele	2012	0.98
AM	BohachevskyIII	888	1.00	AM	Neumaier	14241	0.08
DE	BohachevskyIII	5669	1.00	DE	Neumaier	30661	0.05
GA	BohachevskyIII	28926	0.61	GA	Neumaier	34594	0.20
SA	BohachevskyIII	3393	1.00	SA	Neumaier	46183	0.03
PS	BohachevskyIII	1194	1.00	PS	Neumaier	2514	0.05
QPS	BohachevskyIII	4689	1.00	QPS	Neumaier	10797	0.04
VI	BohachevskyIII	151	1.00	VI	Neumaier	27703	0.09
VII	BohachevskyIII	1482	1.00	VII	Neumaier	14997	0.05
VInD	BohachevskyIII	56	1.00	VInD	Neumaier	44695	0.30
AM	Booth	532	1.00	AM	Paviani	914	1.00
DE	Booth	7776	1.00	DE	Paviani	6979	1.00
GA	Booth	6120	1.00	GA	Paviani	23858	1.00
SA	Booth	1173	1.00	SA	Paviani	NA	0.00
PS	Booth	562	1.00	PS	Paviani	2257	0.91
QPS	Booth	1541	1.00	QPS	Paviani	13333	1.00
VI	Booth	96	1.00	VI	Paviani	NA	0.00
VII	Booth	643	1.00	VII	Paviani	NA	0.00
VInD	Booth	54	1.00	VInD	Paviani	NA	0.00

Algorithm	Problem	Avg	%	Algorithm	Problem	Avg	%
AM	Branin	391	0.99	AM	Periodic	82	0.02
DE	Branin	3690	1.00	DE	Periodic	1918	0.05
GA	Branin	1919	1.00	GA	Periodic	3388	1.00
SA	Branin	1096	1.00	SA	Periodic	517	0.02
PS	Branin	542	1.00	PS	Periodic	4149	0.52
QPS	Branin	658	1.00	QPS	Periodic	11469	0.46
VI	Branin	163	1.00	VI	Periodic	133	0.05
VII	Branin	755	1.00	VII	Periodic	376	0.07
VInD	Branin	191	1.00	VInD	Periodic	77	0.03
AM	Colville	2454	0.91	AM	Perm	275	0.42
DE	Colville	58712	0.22	DE	Perm	1102	0.98
GA	Colville	40829	0.06	GA	Perm	5252	1.00
SA	Colville	56867	0.08	SA	Perm	166	1.00
PS	Colville	4082	0.25	PS	Perm	182	1.00
QPS	Colville	30504	0.04	QPS	Perm	221	1.00
VI	Colville	3587	0.93	VI	Perm	258	0.73
VII	Colville	4922	0.18	VII	Perm	1049	1.00
VInD	Colville	7634	1.00	VInD	Perm	346	0.88
AM	CosineMix	118	1.00	AM	Rastrigin	NA	0.00
DE	CosineMix	287	1.00	DE	Rastrigin	13139	1.00
GA	CosineMix	140	1.00	GA	Rastrigin	23091	1.00
SA	CosineMix	291	1.00	SA	Rastrigin	NA	0.00
PS	CosineMix	145	1.00	PS	Rastrigin	NA	0.00
QPS	CosineMix	138	1.00	QPS	Rastrigin	22196	1.00
VI	CosineMix	60	1.00	VI	Rastrigin	1828	0.00
VII	CosineMix	71	1.00	VII	Rastrigin	6535	0.00
VInD	CosineMix	64	1.00	VInD	Rastrigin	NA	0.00
AM	Dekkers	2003	0.62	AM	Rosenbrock	7409	0.93
DE	Dekkers	5010	0.97	DE	Rosenbrock	39961	0.72
GA	Dekkers	5828	1.00	GA	Rosenbrock	79100	0.00
SA	Dekkers	2583	1.00	SA	Rosenbrock	NA	0.00
PS	Dekkers	3905	0.61	PS	Rosenbrock	3400	0.00
QPS	Dekkers	1203	1.00	QPS	Rosenbrock	54841	0.53
VI	Dekkers	486	0.82	VI	Rosenbrock	57000	0.00
VII	Dekkers	3671	0.66	VII	Rosenbrock	NA	0.00
VInD	Dekkers	594	0.74	VInD	Rosenbrock	NA	0.00

Algorithm	Problem	Avg	%	Algorithm	Problem	Avg	%
AM	Dixon	334	0.37	AM	Salomon	274	0.00
DE	Dixon	6156	0.57	DE	Salomon	71977	0.13
GA	Dixon	21874	0.98	GA	Salomon	57279	0.01
SA	Dixon	1426	0.77	SA	Salomon	NA	0.00
PS	Dixon	615	0.95	PS	Salomon	2104	0.01
QPS	Dixon	1335	0.96	QPS	Salomon	33078	0.11
VI	Dixon	201	0.44	VI	Salomon	1337	0.01
VII	Dixon	657	0.77	VII	Salomon	38777	0.13
VInD	Dixon	381	0.53	VInD	Salomon	61453	0.01
AM	Easom	745	0.91	AM	Schwefel	NA	0.00
DE	Easom	42094	1.00	DE	Schwefel	19759	0.97
GA	Easom	19855	0.46	GA	Schwefel	NA	0.00
SA	Easom	6554	1.00	SA	Schwefel	NA	0.00
PS	Easom	1240	1.00	PS	Schwefel	NA	0.00
QPS	Easom	16638	0.88	QPS	Schwefel	NA	0.00
VI	Easom	981	1.00	VI	Schwefel	NA	0.00
VII	Easom	1793	1.00	VII	Schwefel	NA	0.00
VInD	Easom	NA	0.00	VInD	Schwefel	NA	0.00
AM	Exponential	327	0.95	AM	Shekel	547	0.83
DE	Exponential	2710	1.00	DE	Shekel	16245	1.00
GA	Exponential	1852	1.00	GA	Shekel	7099	0.39
SA	Exponential	52123	0.02	SA	Shekel	55765	0.16
PS	Exponential	527	0.43	PS	Shekel	1084	0.50
QPS	Exponential	3206	1.00	QPS	Shekel	3054	1.00
VI	Exponential	211	0.93	VI	Shekel	2261	0.78
VII	Exponential	422	0.99	VII	Shekel	4583	0.14
VInD	Exponential	1573	1.00	VInD	Shekel	3079	1.00
AM	Goldstein	144	1.00	AM	Sphere	638	1.00
DE	Goldstein	2094	1.00	DE	Sphere	6589	1.00
GA	Goldstein	722	1.00	GA	Sphere	32659	1.00
SA	Goldstein	599	1.00	SA	Sphere	NA	0.00
PS	Goldstein	251	1.00	PS	Sphere	1765	1.00
QPS	Goldstein	342	1.00	QPS	Sphere	11123	1.00
VI	Goldstein	141	1.00	VI	Sphere	50517	0.05
VII	Goldstein	407	1.00	VII	Sphere	4625	0.09
VInD	Goldstein	157	1.00	VInD	Sphere	106	1.00

Algorithm	Problem	Avg	%	Algorithm	Problem	Avg	%
AM	Griewank	2713	0.63	AM	Step	NA	0.00
DE	Griewank	24038	1.00	DE	Step	2803	1.00
GA	Griewank	NA	0.00	GA	Step	7961	1.00
SA	Griewank	NA	0.00	SA	Step	1183	1.00
PS	Griewank	NA	0.00	PS	Step	394	1.00
QPS	Griewank	NA	0.00	QPS	Step	3019	1.00
VI	Griewank	NA	0.00	VI	Step	536	0.77
VII	Griewank	NA	0.00	VII	Step	4398	0.98
VInD	Griewank	803	0.27	VInD	Step	NA	0.00
AM	HelicalValley	759	0.06	AM	SumSquares	769	1.00
DE	HelicalValley	2513	0.00	DE	SumSquares	7758	1.00
GA	HelicalValley	10707	1.00	GA	SumSquares	61844	0.97
SA	HelicalValley	4108	0.07	SA	SumSquares	NA	0.00
PS	HelicalValley	1065	0.01	PS	SumSquares	3081	0.40
QPS	HelicalValley	3375	0.43	QPS	SumSquares	14169	1.00
VI	HelicalValley	372	0.24	VI	SumSquares	74454	0.01
VII	HelicalValley	1086	0.27	VII	SumSquares	6271	0.01
VInD	HelicalValley	457	0.02	VInD	SumSquares	106	1.00
AM	Hosaki	207	0.93	AM	Trid	429	1.00
DE	Hosaki	1033	1.00	DE	Trid	2243	1.00
GA	Hosaki	470	1.00	GA	Trid	1227	1.00
SA	Hosaki	662	1.00	SA	Trid	762	1.00
PS	Hosaki	336	1.00	PS	Trid	360	1.00
QPS	Hosaki	283	1.00	QPS	Trid	498	1.00
VI	Hosaki	147	1.00	VI	Trid	81	1.00
VII	Hosaki	610	1.00	VII	Trid	389	1.00
VInD	Hosaki	180	0.99	VInD	Trid	54	1.00
AM	Himmelblau	394	0.60	AM	Whitley	NA	0.00
DE	Himmelblau	5670	0.99	DE	Whitley	32149	0.59
GA	Himmelblau	2642	1.00	GA	Whitley	83214	0.02
SA	Himmelblau	1341	1.00	SA	Whitley	NA	0.00
PS	Himmelblau	929	0.98	PS	Whitley	NA	0.00
QPS	Himmelblau	716	1.00	QPS	Whitley	26593	1.00
VI	Himmelblau	607	0.89	VI	Whitley	NA	0.00
VII	Himmelblau	2564	0.94	VII	Whitley	NA	0.00
VInD	Himmelblau	584	0.98	VInD	Whitley	NA	0.00

Algorithm	Problem	Avg	%	Algorithm	Problem	Avg	%
AM	Humpl	269	0.87	AM	Zakharov	1577	1.00
DE	Humpl	2085	1.00	DE	Zakharov	NA	0.00
GA	Humpl	1864	1.00	GA	Zakharov	NA	0.00
SA	Humpl	856	1.00	SA	Zakharov	NA	0.00
PS	Humpl	439	1.00	PS	Zakharov	5700	0.00
QPS	Humpl	500	1.00	QPS	Zakharov	NA	0.00
VI	Humpl	274	0.98	VI	Zakharov	NA	0.00
VII	Humpl	753	0.98	VII	Zakharov	NA	0.00
VInD	Humpl	260	1.00	VInD	Zakharov	NA	0.00

4 Multiobjective Optimization

4.1 Comparable and Incomparable Points

Multiobjective optimization attempts to optimize two or more dependent variables simultaneously. This is quite different than the single objective optimization problems discussed thus far.

In single objective optimization, there is only one objective function. The objective function may be difficult to evaluate, nondifferentiable, or resident in a high-dimensional space. However, given a particular point in space, $\vec{x}$, the objective function results in a single value.

In principle, we can evaluate the objective function at every point in space and simply find the largest value. This is the global maximum for the objective function. This may be computationally intractable in practice, but in principle, some global maximum exists.

Fundamentally, the property that a maximum or minimum exists rests on the fact that two different values of the objective function may be compared and rank ordered. If we compute the objective function at one point and find the value 2, then compute at another point to find the value 1, we can say that the value at the first point is greater than the value at the second point. We can do this for any two values of the objective function we come up with. This property means that any pair of values for the objective function may be compared with each other and rank ordered.

This is not true in the multiobjective case. Multiobjective functions produce two or more values from a single point in space. We demonstrate this property with the case of two dependent variables.

Suppose we have a multiobjective function with two dependent variables. Say we are examining a manufacturing plant and are looking to optimize quality and efficiency. We have created some model for each of these dependent variables, and have identified some set of independent variables we can plug into the model. With any set of dependent variables, we can compute a value for quality and a value for efficiency. We want both of these to be as high as possible.

Suppose under one set of independent variable values, we find the quality (Q) has the value .5, and the efficiency (E) has the value .8. Under a different set of independent variable values we find $Q = .6$ and $E = .9$. We would consider the second point preferable over the first point because the second point has a higher value for both Q and E.

Next, suppose the second point has $Q = .6$ and $E = .9$, and a third point has $Q = .8$ and $E = .7$. In this case, the third point has a better value for Q, but a worse value for E. There is no immediate way to determine which of these points is 'better'. Such points are called incomparable because there is no method to rank order these points relative to each other.

Figure 54 shows comparable and incomparable regions for a given point. If the axes are arranged where 'good' is increasing along the positive directions of the axes, then points to the upper right of a test point are always better, while points to the lower left are always worse. Points in other regions (upper left and lower right) are incomparable.

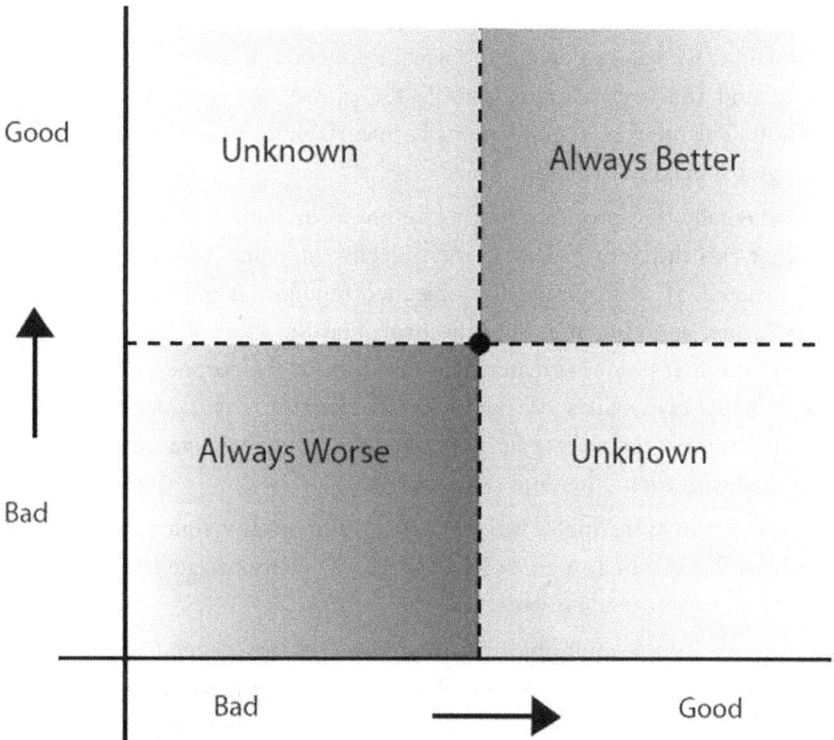

Figure 54: Consider the point marked where the axes are oriented so that 'better' values are to the right along the x-axis and up in the y-axis. Relative to the point in question, the shaded area to the right and up are points that are considered definitively better as all of these points have values for both x and y. The region to the left and below are considered definitively worse because all of these points have lesser values for both x and y. The regions marked 'Unknown' are not comparable to the point in question. Points in this region are better along one axis while worse on the other.

Which regions are better or worse depends on the orientation of the axes. For example, if we plot 'cost' along the x-axis with increasing cost as higher values, then the axis is ordered from good to bad instead of bad to good. Here, low cost (good) is at the origin, while high cost (bad) increases in the positive x direction.

Figure 55 shows the comparable/incomparable regions for the four possible orientations of the axes in two dimensions. Given a point, we can label each region with two letters: G for 'Good' and B for 'Bad'. Each region has two labels, one for each axes. When a region has two G's, the region is always better. When a region has two B's, the region is always worse. Mixed regions are incomparable.

Figure 55 also indicated the labeling of each region with two letter combinations. Using this, we can identify the Good-Bad-Unknown regions for each orientation.

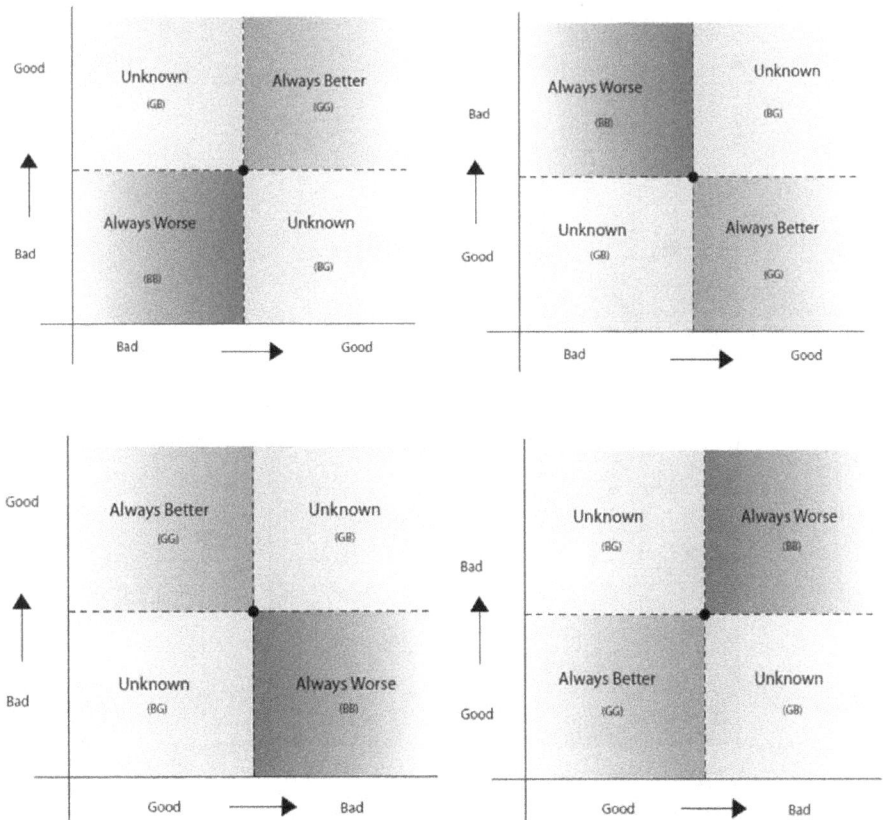

Figure 55: These are the four orientations of the axes in two dimensions. Each diagram presents the regions for Pareto-superior (Always Better), Pareto-inferior (Always Worse), and Pareto incomparable (Unknown) as determined by the orientation of the axes.

This can be extended to higher dimensions. A region that has all G's is always better, the region that has all B's is always worse, and mixed regions are incomparable. In N dimensions, there are 2^N total regions. There will be one unique region of all G's, one unique region for all B's. The remaining $2^N - 2$ regions are incomparable.

When comparing two points, if one point is definitively better than the other, we say that the better point dominates the other. Alternatively, we may describe points as 'Pareto-superior', 'Pareto-inferior', 'Pareto-comparable', or 'Pareto-incomparable'.

4.2 Pareto Frontier

Given a set of points and the orientation (good-bad direction) for each of the axes, we eliminate all points that are Pareto-inferior to at least one other point. The result is a set of points that are not dominated by any other point. These points form the Pareto frontier.

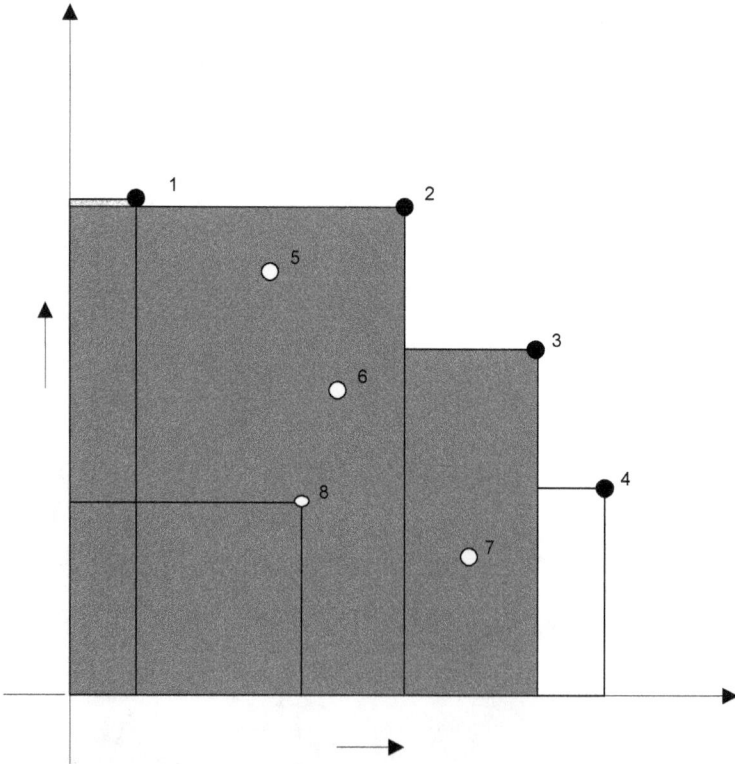

Figure 56: Pareto frontier on eight points. Points 1-4 are on the Pareto frontier, while points 5-8 are not. Points 5-8 are all dominated by at least one other point, while points 1-4 are not dominated by any point.

Figure 56 shows an example of the Pareto frontier. Here we have eight points. Points 1-4 are not dominated by any other point, while points 5-8 are each dominated by at least one other point. Points that are not dominated by any other point lie on the Pareto frontier.

The Pareto frontier is useful to eliminate points that are not optimal solutions. If a point is dominated by another point, we would always choose the dominant point over the dominated point. We can eliminate dominated points from consideration when looking for optimal solutions.

The points on the frontier are candidates as optimal solutions. Again, we cannot choose a particular point on the frontier as the optimum. However, we know that the optimum must be on the frontier.

Additional information is required in order to choose a single optimum from the frontier. Tradeoff functions are a common method for choosing an optimum from the frontier. Tradeoff functions assign a single value for the tradeoff of one variable for another. For example, we might determine we are willing to pay $10 for every unit of increase in accuracy. Based on this, we can evaluate the points on the frontier to determine which provided the optimal value. Choose any pair of points and compute

$$V = C_1 - C_2 + 10(A_1 - A_2)$$
4.1

where C represents the cost value and A is the accuracy value. If this expression is positive, then the first point is superior. When the expression is negative, the second point is superior. If the expression is exactly zero, then the points are equivalent under this tradeoff. Tradeoff functions are discussed in more detail in the next chapter.

Tradeoff functions make a multiobjective optimization problem into a single objective optimization problem. The tradeoff function is a single valued function we attempt to optimize.

In higher dimensions, the tradeoff function may apply to only some variables and not others. In this case, the tradeoff function may be useful to eliminate some of the variables but we may not be left with a single optimization problem. In this case, we still have a multiobjective optimization problem, but we are able to reduce the number of dimensions.

We may receive external input from other sources. Given a list of points on the frontier and the tradeoffs between each, we may simply choose a point suitable to taste. This is hardly a mathematical process, and the result is not repeatable. However, in many cases, this method presents a pleasant solution as we are able to combine a mathematical optimization (finding the frontier) with a nonmathematical approach (considering individual taste). This also provides a degree of control over the final solution to the people who will employ the solution to solve a real-world problem.

4.3 Posets and Lattice Theory

A poset is a countable set of points that have a partial ordering. Let a and b be two members of a set. A partial ordering is a binary relation with the properties

reflexivity $a \leqslant a$ 4.2

antisymmetry $a \leqslant b \wedge b \leqslant a \Rightarrow a \asymp b$ 4.3

transitivity $a \leqslant b \wedge b \leqslant c \Rightarrow a \leqslant c$ 4.4

where $\leqslant$ is analogous to $\leq$, $\asymp$ is analogous to $=$, $\wedge$ is a logical 'and', and $\Rightarrow$ is the logical implication (if a then b).

Reflexivity

The reflexivity relation tells us that a point must be less than or equal to itself. We would expect that a point is always equal to itself, and in fact we will see that this relation combined with the antisymmetry relation prove just that.

Antisymmetry

The antisymmetry relation says that is a is less than or equal to b, and b is less than or equal to a, then a and b must be equal. If we replace b with a, we find that a is equivalent to itself.

Transitivity

Transitivity says that is a is less than or equal to b, and b is less than or equal to c, then a must be less than or equal to c.

These are properties we are accustomed to from integers and real numbers. However, other objects may satisfy these and demonstrate some additional properties. In fact, the Pareto-ordering we have already encountered obeys exactly these properties.

Let $\vec{a}$, $\vec{b}$, and $\vec{c}$ be points in $\mathbb{R}^N$. Let the operator $\asymp$ be defined as

$$\vec{a} \asymp \vec{b} \Leftrightarrow a_i = b_i \ \forall \ i = 1..N \qquad 4.5$$

The operator $\asymp$ means that each individual element of two vectors are equivalent. This is the traditional sense of vector equivalence. We use the double headed arrow '$\Leftrightarrow$' to mean that if we wtrite '$\vec{a} \asymp \vec{b}$' then it is also true that '$a_i = b_i \ \forall \ i = 1..N$'. Moreover, if $a_i = b_i \ \forall \ i = 1..N$' then we aslo have '$\vec{a} \asymp \vec{b}$'.

Furthermore, let $\preccurlyeq$ be defined as

$$\vec{a} \preccurlyeq \vec{b} \Leftrightarrow a_i \leq b_i \ \forall \ i = 1..N \qquad\qquad 4.6$$

First, we show that these operators agree with Pareto-comparisons. Second, we show that these operators create a binary relation satisfying reflexivity, antisymmetry, and transitivity.

Pareto-Equivalence and Pareto-Inferior

Figure 6 shows the regions of comparable and incomparable points relative to a test point. Let the test point be $\vec{a}$ where the components of the vector are the values of the point for the particular aspect. For example, if we are analyzing Quality (Q) and Efficiency (E), and a particular point has $Q = .6$ and $E = .9$, then in vector form this is $\vec{a} = (.6, .9)$.

If two points are considered equal if they represent the same point in the space. Thus, if we have two vectors $\vec{a}$ and $\vec{b}$, every component of these vectors must be identical in order for the points to be considered equal. Mathematically,

$$\vec{a} = \vec{b} \Leftrightarrow a_i = b_i \ \forall \ i = 1..N \qquad\qquad 4.7$$

This expression has the same form as equation 4.5. The symbol '=' in 4.7 has the same meaning as '$\asymp$' from 4.5.

A point $\vec{a}$ is Pareto-inferrior to a point $\vec{b}$ whenever $\vec{b}$ lies to the upper right of $\vec{a}$ in Figure 6. What about the case where $\vec{b}$ lies exactly on the line dividing the upper right from the upper left of $\vec{a}$? If $\vec{b}$ lies on this line, then the x-component of $\vec{b}$ is equal to the x-component of $\vec{a}$, but the y-component of $\vec{b}$ is greater than the y-component of $\vec{a}$. $\vec{b}$ is still Pareto-superrior to $\vec{a}$ because we would prefer $\vec{b}$ over $\vec{a}$: we get the same value for x, but a better value for y. This is also true when $\vec{b}$ lies on the line dividing the upper right from the lower right. In general,

$$\vec{a} < \vec{b} \Leftrightarrow a_i \leq b_i \ \forall \ i = 1..N, \exists i \mid a_i < b_i \qquad\qquad 4.8$$

This says that if $\vec{a} < \vec{b}$, then every component of $\vec{a}$ must be less than or equal to the corresponding component of $\vec{b}$, and there is at least one component of $\vec{a}$ that is strictly less than its corresponding component in $\vec{b}$. In fact, the only reason we need the second statement, 'there is at least one component of $\vec{a}$ that is strictly less than irs corresponding component in $\vec{b}$' is to account for the possibility that every component of $\vec{a}$ is equal to its correspondent in $\vec{b}$. If all components are the same, we would not say that $\vec{a} < \vec{b}$. However, if we examine the concept of 'less than or equal to' we see

$$\vec{a} \leq \vec{b} \Longleftrightarrow a_i \leq b_i \ \forall \ i = 1..N \qquad \text{4.9}$$

But this is the same form for the '$\preccurlyeq$' operator in equation 4.6.

Based on this, if we use a vector form to represent our points, and adopt the definitions from 4.5 and 4.6, the sense of the operators '$\simeq$' and '$\preccurlyeq$' agrees with our expatations for Pareto-equivelance and Pareto-inferrior.

Binary Relation

To show that these operations form a binary relation, we need to show that they obey reflexivity, antisymmetry, and transitivity. We treat these separately below.

In each case, the left side of the expressions from equations 4.24.4 are taken as a given, and we show that the right side must be true as a consequence. If the right side must follow from the left side and the definitions of our operators, then we have established that the expression is true. If all three are true, then our operators form a binary relation.

Reflexivity

Examine a point $\vec{a}$. We know

$$a_i = a_i \ \forall \ i = 1..N \qquad \text{4.10}$$

From 4.5 this means

$$\vec{a} \simeq \vec{a} \qquad \text{4.11}$$

Since all of the components of $\vec{a}$ are the same then it is also true that

$$a_i \leq a_i \ \forall \ i = 1..N \qquad \text{4.12}$$

Thus from 4.6

$$\vec{a} \preccurlyeq \vec{a} \qquad \text{4.13}$$

Examining 4.2, the operator '$\preccurlyeq$' is reflexive.

Antisymmetry

Suppose we have two points, $\vec{a}$ and $\vec{b}$, and suppose that

$$\vec{a} \preccurlyeq \vec{b} \qquad\qquad 4.14$$

and

$$\vec{b} \preccurlyeq \vec{a} \qquad\qquad 4.15$$

If we can show that this means that

$$\vec{a} \asymp \vec{b} \qquad\qquad 4.16$$

then antisymmetry is established.

From 4.14 we know,

$$a_i \le b_i \,\forall\, i = 1..N \qquad\qquad 4.17$$

and from 4.15,

$$b_i \le a_i \,\forall\, i = 1..N \qquad\qquad 4.18$$

The individual components a_i and b_i are real numbers. If a and b are real mumbers, and if $a \le b$ and $b \le a$, then we must have $a = b$. We can draw this conclusion for each component in the vectors. This,

$$a_i = b_i \,\forall\, i = 1..N \qquad\qquad 4.19$$

But this means that

$$\vec{a} \asymp \vec{b} \qquad\qquad 4.20$$

which establishes antisymmetry.

Transitivity

Suppose we have three points, , $\vec{a}$, $\vec{b}$, and $\vec{c}$ where

$$\vec{a} \preccurlyeq \vec{b} \qquad\qquad 4.21$$

and

$$\vec{b} \preccurlyeq \vec{c} \qquad\qquad 4.22$$

In terms of components,

$$a_i \le b_i \,\forall\, i = 1..N \qquad\qquad 4.23$$

$$b_i \le c_i \,\forall\, i = 1..N \qquad\qquad 4.24$$

Again, the components are all real numbers. For real numbers if $a \leq b$ and $b \leq c$ then $a \leq c$. This is true for each set of components. Thus,

$$a_i \leq c_i \ \forall \ i = 1..N \qquad\qquad 4.25$$

But this means

$$\vec{a} \preccurlyeq \vec{c} \qquad\qquad 4.26$$

Putting this together we have

$$\vec{a} \preccurlyeq \vec{b} \wedge \vec{b} \preccurlyeq \vec{c} \Rightarrow a \preccurlyeq c \qquad\qquad 4.27$$

Comparing with 4.4, the operators must be transitive.

Since reflexivity, antisymmetry, and transitivity are all established, the operators '$\preccurlyeq$' and '$\asymp$' create a binary relation. This means that our definition of Pareto-comparison is actually a binary relation. From this, the points we are comparing create a poset (partially ordered set).

In many applications, there is a 'best' and 'worst' possible point. For example, suppose we are examining a process in terms of accuracy and efficiency, where both are measured on the range [0,1]. A given setup of a system has particular values for both accuracy and efficiency.

However, the theoretical best point is the point (1,1). This may not be realistically achievable, but we know that no better point exists. Similarly, the worst possible point is (0,0). This is usually achievable, but even if it is not, we know that no worse point exists.

Suppose we have some set of points under examination, and each axis has a minimum and maximum value. If no point is at the maximum possible value, then we add a point representing the maximum possible value. All other points must be less than this. Similarly, if no point exists at the minimum possible value, we add a point at the minimum. All other points must be greater than this point.

Now our set of points under examination contains exactly one point that is comparable to, and greater than, every other point in the set. Furthermore, we have exactly one point that is comparable to and less than every other point.

A poset that contains a countable number of points, and exactly one point that is greater than every other point, and exactly one point that is less than every other point, is a lattice. If we have a finite set of points to examine, and if the axes all have a finite range of values, then we can identify a corresponding lattice.

As an example, examine the points from left diagram of Figure 57. Suppose the axes each have a maximum and minimum value. We can add points at the minimum and maximum values to obtain the right diagram in Figure 57.

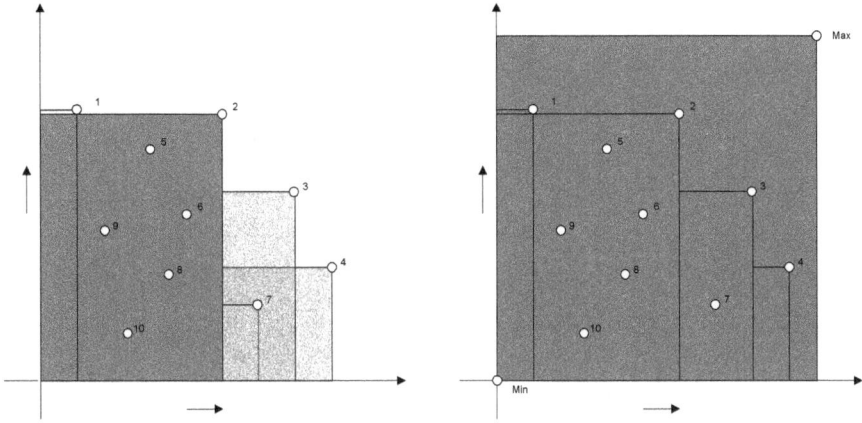

Figure 57: The figure on the left shows a group of points and indicates their Pareto-relationships with boxes. Points falling within the box of another point are Pareto-inferior. The figure on the right shows these same points with additional points representing the Max and Min values.

The points in the right diagram of Figure 57 create a lattice. We know that the points form a poset, and we have a point that is greater than all other points, and a point that is less than all other points.

We can create a diagram for a lattice as well. For a diagram, we put the minimum point at the bottom and the maximum point at the top. The other points are added in between, and we draw a line between two points if one point dominates the other, and if there is no other point that dominates one of these but not the other. This is called a Hasse diagram for the lattice.

Figure 58 shows the Hasse diagram for the points from the right diagram of Figure 57.We place a point for the max at the top of the diagram, and a point for the min at the bottom. Points are joined with a line if one point dominates the other and there are not points 'in between'. For example, point 10 has a line to points 5 and 6, but not to point 2. Point 2 does dominate 10, but both 5 and 6 are between 10 and 2. Thus, we have lines from 5 to 2 and 6 to 2, but not 10 to 2.

By looking at the Hasse diagram we can quickly see how the points are related. The Pareto frontier is the points that directly connect to the Max point. In fact, we see there is a second frontier, formed by points that connect directly to the minimum. These are inferior points, but provide the lower bound for performance under consideration. This lower frontier is the set of points that dominate no point other than the Minimum.

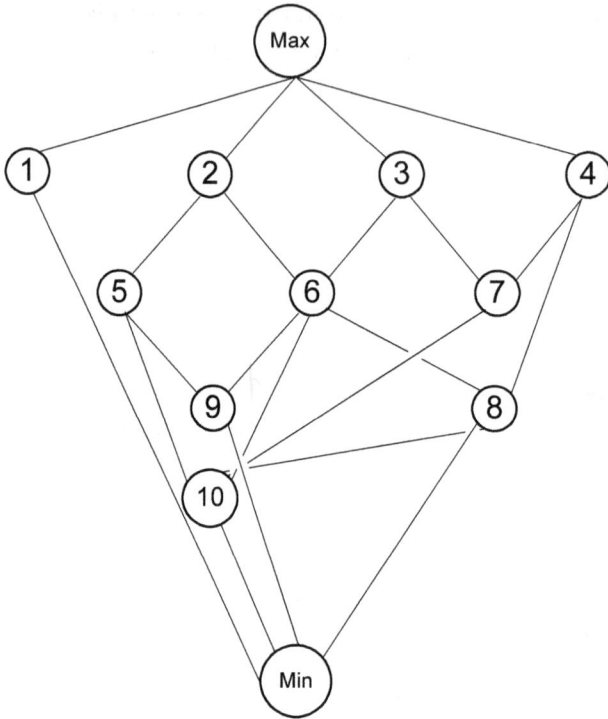

Figure 58: Hasse diagram for the lattice corresponding to the points in Figure 57.

4.4 Dimensional Considerations

As the number of dimensions to the optimization problem increases, the likelihood that a point will be on the frontier also increases. As we add dimensions, the number of ways a point can be incomparable also increases. For example, suppose we have two points to compare. In one dimension, the second point must lie in one of two regions: greater than or less than the first point. In two dimensions, we have four regions relative to the first point: upper right, upper left, lower right, lower left. In n-dimensions, there are 2^n regions. But no matter how many dimensions, there are always only two regions that are Pareto-comparable. The number of incomparable regions is given by

$$R_I(n) = 2^n - 1 \qquad \text{4.28}$$

Because of this, as the number of dimensions increases, the space available for incomparable points also increases. In one dimension there is no incomparable space available. This is the reason we can easily optimize one-dimensional problems: there is no space available for incomparable points, so the maximum must be a unique value.

	5	10	25	50	75	100	250	500	1000
1	0.200	0.100	0.040	0.020	0.013	0.010	0.004	0.002	0.001
2	0.457	0.292	0.153	0.090	0.065	0.052	0.024	0.014	0.007
3	0.668	0.506	0.323	0.218	0.171	0.143	0.077	0.048	0.029
4	0.810	0.686	0.509	0.385	0.320	0.279	0.173	0.117	0.076
5	0.896	0.813	0.671	0.554	0.486	0.439	0.304	0.221	0.158
6	0.945	0.894	0.795	0.700	0.638	0.595	0.453	0.355	0.270
7	0.971	0.942	0.879	0.810	0.762	0.725	0.598	0.497	0.402
8	0.985	0.969	0.932	0.886	0.852	0.825	0.720	0.631	0.537
9	0.992	0.984	0.963	0.935	0.913	0.894	0.817	0.744	0.662
10	0.996	0.992	0.980	0.964	0.950	0.938	0.885	0.833	0.766
11	0.998	0.996	0.990	0.980	0.972	0.965	0.932	0.894	0.845
12	0.999	0.998	0.995	0.990	0.985	0.981	0.961	0.937	0.902
13	0.999	0.999	0.997	0.995	0.992	0.990	0.978	0.963	0.941
14	1.000	0.999	0.999	0.997	0.996	0.995	0.988	0.979	0.965
15	1.000	1.000	0.999	0.999	0.998	0.997	0.994	0.989	0.980
16	1.000	1.000	1.000	0.999	0.999	0.999	0.997	0.994	0.989
17	1.000	1.000	1.000	1.000	0.999	0.999	0.998	0.997	0.994
18	1.000	1.000	1.000	1.000	1.000	1.000	0.999	0.998	0.997
19	1.000	1.000	1.000	1.000	1.000	1.000	1.000	0.999	0.998
20	1.000	1.000	1.000	1.000	1.000	1.000	1.000	1.000	0.999
21	1.000	1.000	1.000	1.000	1.000	1.000	1.000	1.000	1.000
22	1.000	1.000	1.000	1.000	1.000	1.000	1.000	1.000	1.000
23	1.000	1.000	1.000	1.000	1.000	1.000	1.000	1.000	1.000
24	1.000	1.000	1.000	1.000	1.000	1.000	1.000	1.000	1.000
25	1.000	1.000	1.000	1.000	1.000	1.000	1.000	1.000	1.000

Table 1: Probability of a point lying on the Pareto frontier as a function of the dimension and total number of points for Gaussian distributed points with zero mean and unit variance. These values were computed through Monte Carlo simulation. The number of dimensions is on the left and the number of points is across the top.

Table 1 examines the probability that a random Gaussian distributed point lies on the Pareto frontier as the number of dimensions and total number of points varies. If there are only a few points to consider, the probability that the majority of the points lies on the frontier rises quickly with the number of dimensions. For example, in four dimensions and 25 points, we find that nearly half the points lie on the frontier.

This observation provides guidance to multiobjective optimization problems. As the number of objectives increases, the size of the frontier is likely to increase as well. If we attempt to optimize too many simultaneous objectives, we quickly find that nearly every point under consideration is a potential optimum. Having too many dimensions limits the utility of the optimization analysis.

Moreover, if we also consider points that are not on the frontier but are within some error tolerance, we again find the majority of points under consideration as optima. Examining all pairs of such points for their relative tradeoffs can become time consuming and reduce the utility of the optimization process. For these reasons, we need to be careful when choosing objectives to simultaneously optimize.

4.5 Error Analysis

In many cases the numerical values for the objectives along each axis are actually measured values. Measured values can often be associated with a statistical error. An error may be assigned to the performance value along each axis (different errors for different independently measures values).

Knowledge of the error may influence the decision regarding which points to include in the frontier. For example, consider the points $(.5 \pm .2, .5 \pm .1)$ and $(.45 \pm .2, .45 \pm .1)$. The first point dominates the second point with respect to the values: $(.45, .45) \preccurlyeq (.5, .5)$. However, when we account for the errors, the points may be considered equivalent: $(.45 \pm .2, .45 \pm .1) \asymp (.5 \pm .2, .5 \pm .1)$.

In this case, we may decide not to eliminate the second point from the frontier even through it is dominated by the first point. Including the second point on the frontier recognizes that the error bounds are not sufficient to eliminate this as a member of the frontier.

If we include an error component to the analysis, the $\asymp$ may no longer be transitive. For example, if we have a third pint $(.275 \pm .2, .375 \pm .1)$, then we may have $(.275 \pm .2, .375 \pm .1) \asymp (.45 \pm .2, .45 \pm .1)$ and $(.45 \pm .2, .45 \pm .1) \asymp (.5 \pm .2, .5 \pm .1)$, but we may find $(.275 \pm .2, .375 \pm .1) \not\asymp (.5 \pm .2, .5 \pm .1)$.

However, even though the equivalence relation may not be transitive, the $\lessapprox$ can still be transitive. Continuing the example, $(.275 \pm .2, .375 \pm .1) \lessapprox (.45 \pm .2, .45 \pm .1)$ and $(.45 \pm .2, .45 \pm .1) \lessapprox (.5 \pm .2, .5 \pm .1)$, and it is also true that $(.275 \pm .2, .375 \pm .1) \lessapprox (.5 \pm .2, .5 \pm .1)$. The less than or equal to operator is still transitive even though the equivalence operator is not.

Error analysis is useful for comparing two performance points to determine if the points are in fact different or if they are statistically equivalent. This is particularly useful when attempting to optimize a problem by tuning the performance of a solution. For example, we may have an assembly line that makes parts. If we tune the line to make parts faster, the quality may decrease. We may try different techniques to increase the speed while minimizing the impact to quality. Using error analysis, we can determine if these techniques are actually making improvements, or if they are statistically equivalent.

Error analysis is also useful in determining the effective Pareto frontier. Points that are Pareto-inferior, but within error of the frontier, should also be included as part of the frontier. These points are measured as Pareto-inferior, but statistically they are within error tolerance of the frontier. Even though the underlying measurement is Pareto-inferior, these points should be considered as potential optima.

Including points within error on the frontier increases the total number of points under consideration. Coupled with the effect of increasing dimension from the last section, the frontier can quickly contain nearly every point under consideration. If error tolerant points are included on the frontier, it is even more important to limit the number of objectives under consideration in order to provide a useful analysis.

When optimizing several objectives, it may be useful to limit the scope of investigation to just the few objectives that are most interesting. Alternatively, we may find that the values of two objectives are highly correlated. When this occurs, it may be possible to eliminate one of the objectives and only examine the other.

In many cases there are only two dimensions under consideration. With only two dimensions, there are several tools available to compare two points. As the number of dimensions increases, it becomes increasingly difficult to examine the tradeoff between two points.

The next sections examine some techniques for comparing pairs of points. All of these techniques are only useful in two dimensions, others are useful in few dimensions, and some are useful in any number of dimensions. As the number of dimensions increases, the number of analysis approaches available decreases.

4.5-a DISTANCE BETWEEN POINTS

One method to test the difference in the performance metrics of two points is to examine the Euclidean distance between the points relative to the error associated with the distance. The error in the distance may be computed using propagation of errors. We then examine the z-score (ratio of the distance to the square root of the variance) to determine if the distance is statistically significant.

The technique of propagation of errors assumes that the variables are all normally distributed and independent. In many cases, the performance metrics are well approximated as a normal distribution. The assumption here is that the distribution of the distance value is a normally distributed variable.

We begin by examining this in two dimensions. Given a point (χ, φ) and associated variances σ_φ^2 and σ_χ^2, we can identify nearby points that are within a fixed error tolerance. Set τ be a specified error tolerance, and examine the uncertainty in the distance formula between (χ, φ) and another point $(\bar{\chi}, \bar{\varphi})$ with associated errors $\sigma_{\bar{\varphi}}^2$ and $\sigma_{\bar{\chi}}^2$. The distance between these points is

$$d = \sqrt{(\varphi - \bar{\varphi})^2 + (\chi - \bar{\chi})^2} \qquad \text{4.29}$$

The error in this distance may be found through propagation of errors

$$\sigma_d^2 = \left(\frac{\partial d}{\partial \varphi}\right)^2 \sigma_\varphi^2 + \left(\frac{\partial d}{\partial \chi}\right)^2 \sigma_\chi^2 + \left(\frac{\partial d}{\partial \bar{\varphi}}\right)^2 \sigma_{\bar{\varphi}}^2 + \left(\frac{\partial d}{\partial \bar{\chi}}\right)^2 \sigma_{\bar{\chi}}^2 \qquad \text{4.30}$$

$$= \frac{(\varphi - \bar{\varphi})^2 (\sigma_\varphi^2 + \sigma_{\bar{\varphi}}^2) + (\chi - \bar{\chi})^2 (\sigma_\chi^2 + \sigma_{\bar{\chi}}^2)}{d^2} \qquad \text{4.31}$$

Given two different points, one with metrics (χ, φ) and errors $(\sigma_\chi^2, \sigma_\varphi^2)$, and the other with metrics $(\bar{\chi}, \bar{\varphi})$ and errors $(\sigma_{\bar{\chi}}^2, \sigma_{\bar{\varphi}}^2)$, the Euclidean difference in the values is

$$d = \sqrt{(\varphi - \bar{\varphi})^2 + (\chi - \bar{\chi})^2} \qquad \text{4.32}$$

with uncertainty

$$\sigma_d = \frac{\sqrt{(\varphi - \bar{\varphi})^2 (\sigma_\varphi^2 + \sigma_{\bar{\varphi}}^2) + (\chi - \bar{\chi})^2 (\sigma_\chi^2 + \sigma_{\bar{\chi}}^2)}}{d} \qquad \text{4.33}$$

To test if these points are statistically different, we examine the z-score

$$z = \frac{d}{\sigma_d} \qquad \text{4.34}$$

or,

$$z = \frac{(\varphi - \bar{\varphi})^2 + (\chi - \bar{\chi})^2}{\sqrt{(\varphi - \bar{\varphi})^2(\sigma_\varphi^2 + \sigma_{\bar{\varphi}}^2) + (\chi - \bar{\chi})^2(\sigma_\chi^2 + \sigma_{\bar{\chi}}^2)}}$$ 4.35

The z-score is the measure of the statistic in relation to the standard error. This is a two-tailed statistics test, and when the z-score is greater than 1.96, then we are in the 95% confidence interval for statistical significance. A z-score of 1.96 is often used as a measurement of significance. However, other values may be used. In general, the relationship of the z-score to the significance level is

$$S = \sqrt{\frac{2}{\pi}} \int_0^z e^{-\frac{x^2}{2}} dx$$ 4.36

Where S is the significance level and z is the z-score. Table 2 lists some commonly used values of the z-score and the associated significance levels.

	Significance				
z-score	.9	.95	.99	.995	.998
	1.65	1.96	2.58	2.81	3.08

Table 2: z-score v. significance for commonly used statistical tests.

We can use 4.35 to determine the locus of locus of nearby points that all have the same z-score. In this case, we treat $(\bar{\chi}, \bar{\varphi})$ as the measured point with errors $(\sigma_{\bar{\chi}}^2, \sigma_{\bar{\varphi}}^2)$. We want to find all points (χ, φ) that have the same z-score. In this case, the errors on the points (χ, φ) must be $(\sigma_\chi^2 = 0, \sigma_\varphi^2 = 0)$ because these points are exact (they are not measured points, just points on the coordinate system. The expression for the z-score becomes

$$z = \frac{(\varphi - \bar{\varphi})^2 + (\chi - \bar{\chi})^2}{\sqrt{(\varphi - \bar{\varphi})^2\sigma_{\bar{\varphi}}^2 + (\chi - \bar{\chi})^2\sigma_{\bar{\chi}}^2}}$$ 4.37

Multiplying through by the denominator and squaring,

$$z^2[(\varphi - \bar{\varphi})^2\sigma_{\bar{\varphi}}^2 + (\chi - \bar{\chi})^2\sigma_{\bar{\chi}}^2] = [(\varphi - \bar{\varphi})^2 + (\chi - \bar{\chi})^2]^2$$ 4.38

If we make the coordinate change $x = \chi - \bar{\chi}$, $y = \varphi - \bar{\varphi}$,

$$z^2[y^2\sigma_{\bar{\varphi}}^2 + x^2\sigma_{\bar{\chi}}^2] = [y^2 + x^2]^2$$ 4.39

or,

$$(y^2 + x^2)^2 - z^2(\sigma_{\bar{\varphi}}^2 y^2 + \sigma_{\bar{\chi}}^2 x^2) = 0 \qquad\qquad 4.40$$

In general, this is a complicated curve. Figure 59 provides an example of the contours of this curve when x and y have unit error. The inner area is the region of points where z-score is less than 1.96. The points in this region are all within the 95% confidence bound. The outer area is the region of points where z-score is less than 2.58 (less than 99% confidence). The two regions together indicate how large the confidence bounds are and the additional area absorbed as we increase the confidence tolerance.

Figure 60 shows a similar situation where the error of y is twice the error on x. The curve is shaped similar to the intersection of two circles. Comparing with the previous graph, we see that the bounds on the confidence along the x-axis are the same. However, the bounds along the y-axis is twice the previous.

Figure 61 provides an example of how these error bounds might appear relative to a point. We see the same general shape as in Figure 60, however the bounds are centered on a particular performance point. Moreover, Figure 62 shows three points with error bounds. In this case, one of the points is separated from the others. However, two of the points demonstrate some degree of overlap. The 95% regions do not overlap, but the 99% regions do.

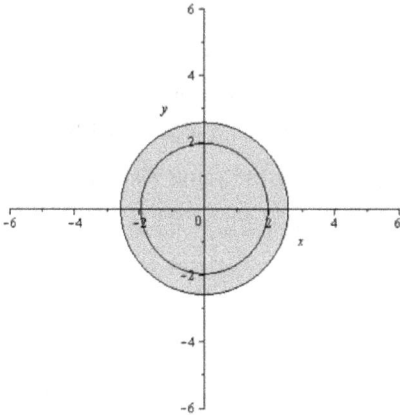

Figure 59: Plot constant z-scores where x and y have unit error. The inner region has z-score < 1.96, while the outer region has z-score < 2.58.

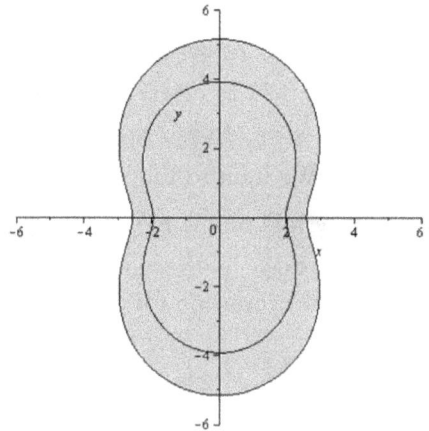

Figure 60: Plot constant z-scores where y has twice the error as x. The inner region has z-score < 1.96, while the outer region has z-score < 2.58.

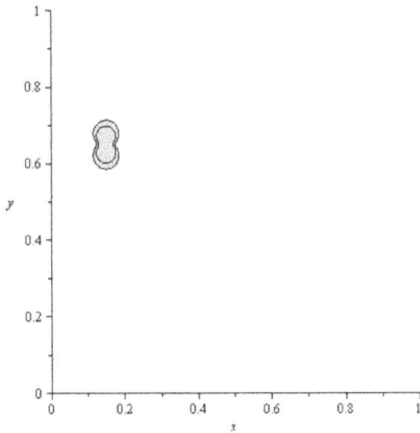

Figure 61: Example plot of metrics with error bound.

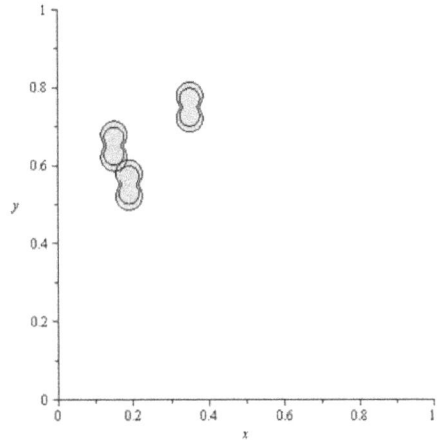

Figure 62: Example plot of multiple points with error bounds.

Overlap of the 99% regions does not mean that these points are insignificantly different at the 99% level. In order to determine if the points are significantly different, we must use 4.37 and compute the z-score for the difference in the points in light of the relative errors. Similarly, if the regions do not overlap, this does not necessarily mean that the points are significantly different. Again, we must use 4.37 to definitively determine the significance.

The error bounds on the graph are meant to provide an indication of the regions of the z-scores. When error bounds highly overlap, this is a good indication that the points are not significantly different. When the error bounds are very far apart, this is a good indication that the points are significantly different. However, in any case, we should evaluate 4.37 to determine the significance.

We can extend this to an arbitrary number of dimensions. In general,

$$z = \frac{\sum_{i=1}^{N}(x_i - \bar{x}_i)^2}{\sqrt{\sum_{i=1}^{N}(x_i - \bar{x}_i)^2 \left(\sigma_{x_i}^2 + \sigma_{\bar{x}_i}^2\right)}}$$

4.41

Examining the Euclidean distance is a natural method for comparing two points. However, the distance can be dominated by the difference in values of just one component. When one component has a large difference, ant the other is small, the Euclidean distance is dominated by the large difference components. If the error associated with the small difference is very small, there may be a statistically significant difference along one component, but this can be lost when added to the large difference in the second component.

4.5-b SUM OF Z-SCORES

The drawback to using the distance as a measure of significance is that if difference in one metric is much smaller than the other metric, then the second difference will dominate the distance. This is even worse when the variance associated with the first distance is much smaller than the second.

Another approach is to examine the sum of the z-scores for each individual metric. Here, each difference is measured relative to its own error, and only the ratio is considered. In this case, if a difference is small, but the variance is much smaller, the z-score is still large. By summing the z-scores we create a statistic that is able to relatively measure each contribution.

The z-score is

$$z = \frac{x}{\sigma_x} \qquad\qquad 4.42$$

where x is a measured value and σ_x^2 is the variance of the measurement. This test examines if the value x is significantly different from 0 (null hypothesis). The z-score is distributed as a Student-t distribution.

Suppose we measure x by sampling. We take multiple independent measurements of x. Let x_i be the ith measurement. The mean and variance are approximated as

$$x = \frac{1}{N} \sum_{i=1}^{N} x_i \qquad\qquad 4.43$$

$$\sigma_x^2 = \frac{1}{N-1} \sum_{i=1}^{N} (x_i - x)^2 \qquad\qquad 4.44$$

If we use these values in equation 4.42, then the z-score is distributed as a Student-t distribution

$$f(t) = \frac{1}{\sqrt{\upsilon} B\left(\frac{1}{2}, \frac{\upsilon}{2}\right)} \left(1 + \frac{t^2}{\upsilon}\right)^{-\frac{\upsilon+1}{2}} \qquad\qquad 4.45$$

where υ is the number of degrees of freedom. If we measure by sampling, $\upsilon = N - 1$. In most cases, we do not measure these by sampling. Instead, we measure the performance metric, then compute the variance by assuming the measurement is distributed according to a normal distribution. This is correct in the limit that we have an infinite number of measurements. Applying this here, we take the limit as $v \to \infty$. In this limit, the distribution becomes a normal distribution with unit variance:

$$\lim_{v \to \infty} f(t) = \frac{1}{\sqrt{2\pi}} e^{-\frac{t^2}{2}} \qquad 4.46$$

For this distribution, the probability that the absolute value of the z-score is greater than or equal to some value t is given by,

$$P(|z| \geq t) = \frac{2}{\sqrt{2\pi}} \int_t^\infty e^{-\frac{z^2}{2}} dz \qquad 4.47$$

$$= \frac{2}{\sqrt{\pi}} \int_{t/\sqrt{2}}^\infty e^{-z^2} dz \qquad 4.48$$

$$P(|z| \geq t) = erfc(t/\sqrt{2}) \qquad 4.49$$

where $erfc(t)$ is the complementary error function

$$erfc(t) = \frac{2}{\sqrt{\pi}} \int_t^\infty e^{-z^2} dz \qquad 4.50$$

It is useful to note

$$\int_0^\infty e^{-z^2} dz = \frac{\sqrt{\pi}}{2} \qquad 4.51$$

We use this relation when evaluating integrals later in this section.

Each z-score has an independent normal distribution. Examine the sum of two z-scores,

$$\bar{K} = x + y = \frac{\varphi - \bar{\varphi}}{\sqrt{\sigma_\varphi^2 + \sigma_{\bar{\varphi}}^2}} + \frac{\chi - \bar{\chi}}{\sqrt{\sigma_\chi^2 + \sigma_{\bar{\chi}}^2}} \qquad 4.52$$

where both x and y are distributed according to 4.46. This sum treats both the x and y variables independently. This statistic is sensitive to both variables and is not dominated by large deviations in one difference relative to the other.

However, this statistic has a different problem. Both x and y can be either positive or negative. We can have a situation where x is large and negative, while y is large and positive, but the sum is zero.

Instead of this statistic, examine

$$K = |x| + |y| \qquad 4.53$$

Now each term is positive so they can't cancel each other out.

Next, we analyze the behavior of this statistic. The variables x and y are independent of each other and normally distributed with unit variance. We can identify the distribution of $|x|$ from 4.46. Examine the statistic $z = |x|$ where x is distributed according to 4.46. We can write z as

$$z = \begin{cases} x & x \geq 0 \\ -x & x < 0 \end{cases} \tag{4.54}$$

Solving this fox x,

$$x = \pm z \tag{4.55}$$

From this, the distribution of z is

$$K_1(z) = \frac{1}{\sqrt{2\pi}} e^{-\frac{z^2}{2}} + \frac{1}{\sqrt{2\pi}} e^{-\frac{(-z)^2}{2}} \tag{4.56}$$

$$= \frac{2}{\sqrt{2\pi}} e^{-\frac{z^2}{2}} \tag{4.57}$$

This distribution is for $z \geq 0$. The full distribution is

$$K_1(z) = \begin{cases} \dfrac{2}{\sqrt{2\pi}} e^{-\frac{z^2}{2}} & z \geq 0 \\ 0 & z < 0 \end{cases} \tag{4.58}$$

With one variable, the probability for $K_1 \geq t$ is

$$P_1(K_1 \geq t) = \frac{2}{\sqrt{2\pi}} \int_t^\infty e^{-\frac{z^2}{2}} dz \tag{4.59}$$

$$= \frac{2}{\sqrt{\pi}} \int_{t/\sqrt{2}}^\infty e^{-z^2} dz \tag{4.60}$$

or

$$P_1(K_1 \geq t) = erfc(t/\sqrt{2}) \tag{4.61}$$

This matches the result from 4.50.

The subscript 1 above indicates we are examining the distribution from one such variable. With two independent variables the distribution of the sum is given by the convolution of the distributions:

$$K_2(z) = h(z) = \int_{-\infty}^\infty K_1(z - w)K_1(w)\, dw \tag{4.62}$$

The integrand vanishes whenever $w < 0$ or when $w > z$. Thus,

$$h(z) = \int_0^z K_1(z-w)K_1(w)\,dw \qquad\qquad 4.63$$

$$= \frac{2}{\pi} \int_0^z e^{-\frac{(z^2-2zw+2w^2)}{2}}\,dw \qquad\qquad 4.64$$

$$= \frac{2}{\pi} e^{-\frac{z^2}{2}} \int_0^z e^{w(z-w)}\,dw \qquad\qquad 4.65$$

$$= \frac{2}{\pi} e^{-\frac{z^2}{4}} \int_0^z e^{-\left(w-\frac{z}{2}\right)^2}\,dw \qquad\qquad 4.66$$

$$= \frac{2}{\pi} e^{-\frac{z^2}{4}} \int_{-\frac{z}{2}}^{\frac{z}{2}} e^{-u^2}\,du \qquad\qquad 4.67$$

$$= \frac{4}{\pi} e^{-\frac{z^2}{4}} \int_0^{\frac{z}{2}} e^{-u^2}\,du \qquad\qquad 4.68$$

$$= \frac{2}{\sqrt{\pi}} e^{-\frac{z^2}{4}} erf(z/2) \qquad\qquad 4.69$$

The probability for $t \le K_2 < \infty$ is

$$P_2(K_2 \ge t) = \frac{2}{\sqrt{\pi}} \int_t^{\infty} e^{-\frac{z^2}{4}} erf(z/2)\,dz \qquad\qquad 4.70$$

$$= \frac{4}{\sqrt{\pi}} \int_{t/2}^{\infty} e^{-z^2} erf(z)\,dz \qquad\qquad 4.71$$

$$= 1 - erf^2(t/2) \qquad\qquad 4.72$$

Extending this statistic to higher dimensions is mathematically difficult. The convolution for higher sums creates increasingly complicated integrals that require numerical techniques to evaluate. This statistic is useful for two dimensions, but is less useful in higher dimensions.

4.5-c   HOTELLING'S T-SQUARE

Instead of using the sum absolute value of the z-score, we could examine the sum of the squares of the z-scores. This approach is similar to Hotelling's T-Square. Hotelling's T-Square applies to a set of measurements of different dependent variables. The technique can be extended to sets with an infinite degree of freedom on multiple dependent variables.

In the previous section, we found that the z-score is distributed as a normal distribution with zero mean and unit variance

$$f(t) = \frac{1}{\sqrt{2\pi}} e^{-\frac{t^2}{2}} \qquad \text{4.73}$$

The distribution of $z = t^2$ is:

$$H_1(z) = \frac{1}{\sqrt{2\pi}} z^{-\frac{1}{2}} e^{-\frac{z}{2}} \qquad \text{4.74}$$

This is a χ^2 distribution on one degree of freedom. The probability that z exceeds a value t is

$$P_1(H_1 \geq t) = \frac{1}{\sqrt{2\pi}} \int_t^\infty z^{-\frac{1}{2}} e^{-\frac{z}{2}} dz \qquad \text{4.75}$$

$$= \frac{1}{\sqrt{2\pi}} \gamma\left(\frac{1}{2}, \frac{x}{2}\right) \qquad \text{4.76}$$

where γ is the lower incomplete gamma function.

The sum of two variables distributed according to 4.74 can be found from the convolution of the distribution:

$$H_2(z) = \int_{-\infty}^\infty H_1(z-w) H_1(w) \, dw \qquad \text{4.77}$$

$$= \frac{1}{2\pi} \int_{-\infty}^\infty (z-w)^{-\frac{1}{2}} e^{-\frac{(z-w)}{2}} w^{-\frac{1}{2}} e^{-\frac{w}{2}} \, dw \qquad \text{4.78}$$

$$= \frac{1}{2\pi} e^{-\frac{z}{2}} \int_{-\infty}^\infty (z-w)^{-\frac{1}{2}} w^{-\frac{1}{2}} \, dw \qquad \text{4.79}$$

$$= \frac{1}{2\pi} z^{-1} e^{-\frac{z}{2}} \int_{-\infty}^\infty \left(1 - \frac{w}{z}\right)^{-\frac{1}{2}} \frac{w^{-\frac{1}{2}}}{z} \, dw \qquad \text{4.80}$$

$$= \frac{1}{2\pi} e^{-\frac{z}{2}} \int_{-\infty}^\infty (1-u)^{-\frac{1}{2}} u^{-\frac{1}{2}} \, du \qquad \text{4.81}$$

$$= \frac{1}{2} e^{-\frac{z}{2}} \qquad \text{4.82}$$

This is also a χ^2 distribution, but this is on two degrees of freedom. The probability that z exceeds a value t is

$$P_2(H_2 \geq t) = \frac{1}{2} \int_t^\infty e^{-\frac{z}{2}} dz \qquad \text{4.83}$$

$$= \gamma\left(1, \frac{t}{2}\right) \qquad \text{4.84}$$

From these results, we may examine the statistic

$$H = x + y = \frac{(\varphi - \bar{\varphi})^2}{\sigma_\varphi^2 + \sigma_{\bar{\varphi}}^2} + \frac{(\chi - \bar{\chi})^2}{\sigma_\chi^2 + \sigma_{\bar{\chi}}^2} \qquad \text{4.85}$$

We compute this value for a particular binary classifier. The confidence level of difference is determined from P_2:

$$P_2(H_2 \geq H) = \gamma\left(1, \frac{H}{2}\right) \qquad \text{4.86}$$

Here, we use the lower incomplete gamma function to determine the probability that we would find a value of H simply by random chance. A table of this function is provided in Appendix E.

The most useful aspect of this approach is that the convolution of two Student-t distributions results in another Student-t distribution. This makes it easy to extend the statistic to higher dimensions. For example, we can compute the distribution for three dimensions by convoluting the two-dimensional distribution with a one-dimensional distribution:

$$H_3(z) = \int_{-\infty}^{\infty} H_2(w) H_1(z - w)\, dw \qquad \text{4.87}$$

$$= \frac{1}{2\sqrt{2\pi}} \int_{-\infty}^{\infty} e^{-\frac{w}{2}} (z - w)^{-\frac{1}{2}} e^{-\frac{(z-w)}{2}}\, dw \qquad \text{4.88}$$

$$= \frac{1}{2\sqrt{2\pi}} z^{-1/2} e^{-\frac{z}{2}} \int_{-\infty}^{\infty} \left(1 - \frac{w}{z}\right)^{-\frac{1}{2}}\, dw \qquad \text{4.89}$$

$$= \frac{1}{2\sqrt{2\pi}} z^{1/2} e^{-\frac{z}{2}} \int_{-\infty}^{\infty} (1 - u)^{-\frac{1}{2}}\, du \qquad \text{4.90}$$

$$= \frac{1}{\sqrt{2\pi}} z^{1/2} e^{-\frac{z}{2}} \qquad \text{4.91}$$

This is again a χ^2 distribution. In this case the distribution is on three degrees of freedom. The probability that the statistic is larger than some value H by chance is

$$P_3(H_3 \geq H) = \frac{2}{\sqrt{\pi}} \gamma \left(\frac{3}{2}, \frac{H}{2} \right) \tag{4.92}$$

$$= P \left(\frac{3}{2}, \frac{H}{2} \right) \tag{4.93}$$

where $P(n, x)$ is the regularized gamma distribution. This leads us to suspect that the sum of n such variables might be a χ^2 distribution with n-degrees of freedom. We prove this is true by induction.

The χ^2 distribution on n-degrees of freedom is

$$\chi^2(n; z) = \frac{1}{2^{n/2} \Gamma(n/2)} z^{\frac{n}{2}-1} e^{-\frac{z}{2}} \tag{4.94}$$

Suppose

$$H_n(z) = \chi^2(n; z) \tag{4.95}$$

Then the sum on $n + 1$ such variables is given by the convolution of this with H_1

$$H_{n+1}(z) = \int_{-\infty}^{\infty} H_n(w) H_1(z - w) \, dw \tag{4.96}$$

$$= \frac{1}{2^{n/2} \sqrt{2\pi} \Gamma(n/2)} \int_{-\infty}^{\infty} w^{\frac{n}{2}-1} e^{-\frac{w}{2}} (z - w)^{-\frac{1}{2}} e^{-\frac{(z-w)}{2}} \, dw \tag{4.97}$$

$$= \frac{1}{2^{n/2} \sqrt{2\pi} \Gamma(n/2)} e^{-\frac{z}{2}} \int_{-\infty}^{\infty} w^{\frac{n}{2}-1} (z - w)^{-\frac{1}{2}} \, dw \tag{4.98}$$

$$= \frac{1}{2^{n/2} \sqrt{2\pi} \Gamma(n/2)} z^{-\frac{1}{2}} e^{-\frac{z}{2}} \int_{-\infty}^{\infty} w^{\frac{n}{2}-1} \left(1 - \frac{w}{z} \right)^{-\frac{1}{2}} \, dw \tag{4.99}$$

$$= \frac{1}{2^{n/2} \sqrt{2\pi} \Gamma(n/2)} z^{\frac{n-3}{2}} e^{-\frac{z}{2}} \int_{-\infty}^{\infty} \left(\frac{w}{z} \right)^{\frac{n}{2}-1} \left(1 - \frac{w}{z} \right)^{-\frac{1}{2}} \, dw \tag{4.100}$$

$$= \frac{1}{2^{n/2} \sqrt{2\pi} \Gamma(n/2)} z^{\frac{n-1}{2}} e^{-\frac{z}{2}} \int_{-\infty}^{\infty} u^{\frac{n}{2}-1} (1 - u)^{-\frac{1}{2}} \, du \tag{4.101}$$

$$= \frac{1}{2^{(n+1)/2} \Gamma((n + 1)/2)} z^{\frac{n+1}{2}-1} e^{-\frac{z}{2}} \tag{4.102}$$

$$= \chi^2(n + 1; z) \tag{4.103}$$

Thus, if H_n is a χ^2 distribution on n-degrees of freedom, then H_{n+1} is a χ^2 distribution on $n + 1$-degrees of freedom. Since we know that H_1 is a χ^2 distribution with 1 degree of freedom, the proof by induction is complete.

From this, the probability that the statistic is larger than H by chance is

$$P_n(H_n \geq H) = \frac{1}{\Gamma\left(\frac{n}{2}\right)} \gamma\left(\frac{n}{2}, \frac{x}{2}\right) \qquad 4.104$$

$$= P\left(\frac{n}{2}, \frac{x}{2}\right) \qquad 4.105$$

where $P(n, x)$ is the regularized gamma function.

The Hotelling statistic is useful when comparing points in any number of dimensions. In general, the statistic is

$$H = \sum_{i=1}^{N} \frac{(x_i - \bar{x}_i)^2}{\sigma_{x_i}^2 + \sigma_{\bar{x}_i}^2} \qquad 4.106$$

The statistic is easy to compute and may be compared with tables of the regularized gamma function to determine the level of significance.

The Hotelling statistic determines if two points are statistically different. It does not tell us which is better, only that the two are different. When assessing performance, we use this test in conjunction with the others in this section to identify exactly how two classifiers compare. Alternatively, if we are tuning performance, this statistic is very useful for determining when the tuning has a statistically significant effect on performance.

4.5-d WEIGHTED SUMS

Another approach is to compute the weighted sum of the terms from the Hotelling statistic. In some cases there may be a desire to place more emphasis on some variables over others. This should not be confused with a tradeoff analysis (tradeoffs are discussed in later chapters).

The weighted Hotelling statistic is

$$H = \sum_{i=1}^{N} \rho_i \frac{(x_i - \bar{x}_i)^2}{\sigma_{x_i}^2 + \sigma_{\bar{x}_i}^2} \qquad 4.107$$

where the ρ_i's are set using by the analyst.

Weighting the Hotelling statistic alters the significance levels for statistical significance. These must be recomputed using the weights used to calculate the statistic.

4.5-e   METRIC SIGNIFICANCE

Another alternative is to compute the significance between each pair of components in the points and count the number of times one point significantly outperforms the other. This leads to a table for comparing the two points be observing which metrics are significantly better for the first point versus the second point.

Table 3 provides an example of a Metric Significance table. In this example there are five dimensions. Using the z-score, we examine each dimension individually and determine if the first point (A) is better, if the second point (B) is better, or if (N)either are better.

The Metric Significance table is useful for comparing the various aspects of the two points. In the example, we see that A is better in dimensions 1 and 5, while B is better in dimensions 3 and 4. These points must be Pareto-incomparable because neither point is inferior to the other in every dimension.

This table may be used to choose one of these points over the other in specific circumstances. For example, if we are applying optimization to a problem where we have little interest in dimensions 1 and 5, then we might conclude B is the better choice. Similarly, other circumstances may lead to A as the better choice.

1	2	3	4	5
A	N	B	B	A

Table 3: Sample Metric Significance table. In this example there are five dimensions shows in the first row. The second row states if point (A) is significantly better than B, (B) is significantly better that A, or (N)either is significantly better.

4.6 Optimizing Multiple Objectives

This section summarizes the approach outlined in the previous sections for identifying potential optima under multiobjective optimization. We assume that the number of dimensions are determined, and a set of points representing achievable outcomes are provided.

Pareto Frontier

The first step is to identify the points that lie on the Pareto frontier. These are the points that are not dominated by any other point. This frontier is important because the optimum solution must be an achievable point on this frontier.

The frontier may be determined by comparing each point with every other point and discarding every point that is dominated by another point.

Alternatively, we may construct a Hasse diagram and identify the frontier as the points that directly connect to the maximum.

Error Analysis

Error analysis involved comparing pairs of points to determine if they are within error tolerance of each other. Dominated points may lie within error tolerance of a frontier point. In this case, it is useful to include these points as frontier points rather than discarding them as inferior.

Tradeoff Analysis

Once the frontier is identified, we can examine the tradeoff between each pair of points. Typically it is not necessary to examine every pair. In two dimensions, the frontier points can be ordered with respect to one of the dimensions. Once the points are ordered, we only need to compare pairs of points.

In general, we can order the points along any one dimension. We then compare neighboring points and determine the tradeoff across the various dimensions. This is useful when there is a principle dimension that stands out as significant.

No Unique Optimum

In most cases, there is no unique optimal solution for multiobjective optimization problems. We arrive at a set of frontier points that represent the potential optimal solutions. Choosing between these points is often a matter of taste.

Using the frontier to eliminate suboptimal solutions and choosing an optimum among the frontier points provides an opportunity for the user to exercise a degree of control over the final solution. This approach is often a powerful combination between a mathematical solution from the frontier combined with the control to choose a preferred operating point by a user.

5 Tradeoff Functions

5.1 Rational Tradeoff Functions

Tradeoff functions are functions of the dependent variables that specify the relative value of the dependent variables. For example, we might be willing to give up some production efficiency to gain more quality. A tradeoff function may be

$$Q = -2E \qquad\qquad 5.1$$

This means that we are willing to give up two units of quality to gain one unit of efficiency.

Tradeoff functions may be used to decide between points on the frontier. Continuing with the example tradeoff function, suppose we have the three point frontier shown in Figure 63. These points are incomparable because no point is dominated by either of the other two.

The tradeoff function may be used to create a value function. The tradeoff function determines how much of one variable we are willing to give up to receive benefit in another variable. Our function may be rewritten as

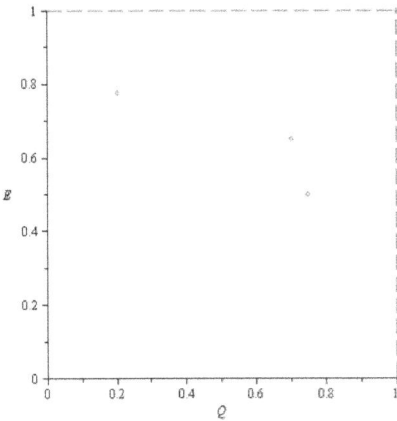

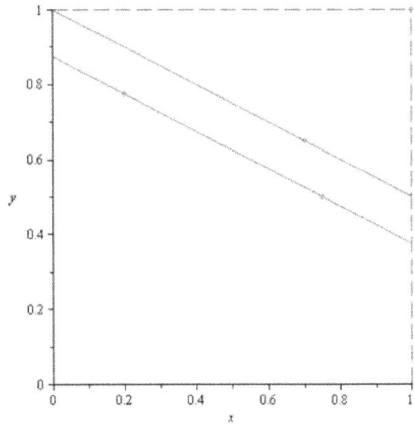

Figure 63: Sample optimization frontier examining (E)fficiency and (Q)uality.

Figure 64: Frontier with lines of constant value.

$$Q + 2E = 0 \qquad \text{5.2}$$

The tradeoff function examines the relative value of two points. This can be used to create value curves of the form

$$Q + 2E = \gamma \qquad \text{5.3}$$

where γ represents the overall value. In the value function, we can substitute the variables for a single point to arrive at an overall value for the point. Higher values are considered 'better' than lower values.

If we do this with the points in Figure 63 we arrive at the values in Table 4. Two of the points have the same value. This means that the points are considered equivalent under the tradeoff/value functions. The third point has a higher value. This point is preferred over the other two according to the tradeoff/value functions.

Q	E	Value
.750	.500	1.75
.200	.775	1.75
.700	.650	2.00

Table 4: Values for the points from Figure 63 using equation 5.3.

Figure 64 shows the points with lines of constant value. One of the lines of constant value goes through two points, while the third only intersects one of the points. Again, we see that the first two points in Table 4 have the same value according to the tradeoff function, while the third point has a higher value.

Tradeoff functions are used to rank order the points on the Pareto frontier to determine a unique optimal solution. Essentially, tradeoff functions provide a single-dependent variable that may be optimized using traditional optimization methods.

In this chapter we assume that the dependent variables are ordered in a manner so that 'better' is in the direction of increasing value for the variable. This corresponds to the axes shown in Figure 54 of the previous chapter. The methodologies we present work for other orientations, but must be altered appropriately.

In principal, any function of the multiple dependent variables can be used as a tradeoff function. However, many functions do not make sense in terms of multiobjective optimization. For example, the tradeoff function

$$Q = 2E \qquad \text{5.4}$$

would mean that we are willing to give up one unit of quality if we also give up two units of efficiency. This makes no sense as a tradeoff function. We don't want to give up either quality or efficiency, and we certainly do not want to dive up both at the same time.

Irrational tradeoffs such as this are called pathological tradeoffs. Pathological tradeoff functions connect points that are comparable. Because of this, we receive two pieces on conflicting information. First, since the points are Pareto comparable, we know that one point is considered superior to the other. Second, the tradeoff functions tell us that the points are considered the same. We discuss pathological tradeoff functions in more detail in the next section.

Based on these examples, we can generalize how tradeoff and value functions are expected to behave. We begin by examining lines connecting incomparable points in two dimensions, and then extend those results to higher dimensions.

In two dimensions, if the axes are oriented so that 'better' is in the positive direction, then comparable points are connected by non-negatively sloped lines. Lines may be positively sloped, have zero slope, or infinite slope. Lines connecting two points with these slopes connect points that are comparable. Incomparable points are connected only by negatively sloped lines. A line connecting two points in two dimensions has the form

$$\alpha x + \beta y + \gamma = 0 \qquad \text{5.5}$$

Taking the differential

$$\alpha dx + \beta dy = 0 \qquad \text{5.6}$$

The slope is the ratio of the differentials

$$\frac{dx}{dy} = -\frac{\beta}{\alpha} \qquad \text{5.7}$$

or

$$\frac{dy}{dx} = -\frac{\alpha}{\beta} \qquad \text{5.8}$$

There are two slopes we can consider. The slope corresponding to changes in x with respect to changes in y (5.7), and the slope corresponding to changes in y with respect to changes in x (5.8). Although these slopes have different numerical values, the sign of the slope is the same in both cases.

When $\alpha > 0$ and $\beta > 0$, these slopes are negative, and the line connects incomparable points. When either $\alpha \leq 0$ and $\beta > 0$, or $\alpha > 0$ and $\beta \leq 0$, the line is non-negatively sloped and connects comparable points.

In higher dimensions, lines take the form

$$\sum_{k=1}^{N} \alpha_k x_k + \gamma = 0 \tag{5.9}$$

where the x_k's are the dependent variables. Taking the differential,

$$\sum_{k=1}^{N} \alpha_k dx_k = 0 \tag{5.10}$$

If the coefficients satisfy $\alpha_k > 0 \ \forall \ k = 1, .., N$, then the line connects incomparable points. Otherwise the line connects comparable points.

Suppose we have a tradeoff function of the form

$$f(\vec{x}) = 0 \tag{5.11}$$

where $\vec{x}$ is the vector of dependent variables we are optimizing over. We can convert this to a value function by setting

$$f(\vec{x}) = \gamma \tag{5.12}$$

where γ is the value we associate with any particular $\vec{x}$, and higher values of γ indicate better value. If we compute the total derivative,

$$df = \sum_{k=1}^{N} \frac{\partial f(\vec{x})}{\partial x_k} dx_k = 0 \tag{5.13}$$

At a given point $\vec{x}$, this takes the form of 5.10 where

$$\alpha_k = \frac{\partial f(\vec{x})}{\partial x_k} \tag{5.14}$$

The tradeoff function is rational if the partial derivative with respect to each variable is positive at every point. Based on this, we say that a tradeoff function $f(\vec{x})$ is rational if

$$\alpha_k = \frac{\partial f(\vec{x})}{\partial x_k} > 0 \quad k = 1 \dots N \tag{5.15}$$

Any proposed tradeoff function that does not meet this condition is considered a pathological tradeoff function.

Rational tradeoff functions lead to rational value curves. The family of value curves generated from a rational tradeoff function also have the same property as in equation 5.15.

The family of value curves obeying 5.15 do not cross. To prove this, suppose that two value curves cross at some point. This situation is shown in Figure 65 near the crossing point. Both of the value curves are negatively sloped as required, and cross at some point. The points on one of these lines are considered equivalent. But since the lines cross, the value assigned to each line must be the same. Thus, the points on both lines are equivalent. Since the lines cross, there must be a point on one line that dominates at least one point on the other line. In one sense, these points are equivalent because they have the same value. In the Pareto sense, one dominates the other. This is a contradiction, so our initial assumption that value curves cross must be incorrect.

Figure 65: Rational value curves cannot cross because this would mean that some pair of points are both equivalent and Pareto comparable.

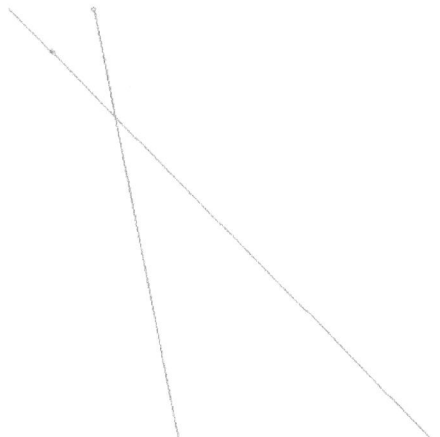

Figure 66: If curves of constant value cross, there are two points that are equivalent by value but one point dominates the other.

For a more complicated example suppose we have the family of value curves

$$f(Q, E) = Q + 2E^2 = \gamma \qquad \text{5.16}$$

where $0 \leq Q \leq 1$ and $0 \leq E \leq 1$. Checking the partial derivatives,

$$\frac{\partial f(Q, E)}{\partial Q} = 1 \qquad \text{5.17}$$

$$\frac{\partial f(Q, E)}{\partial E} = 2E \qquad \text{5.18}$$

Both partial derivatives are positive everywhere that $0 < E \leq 1$. At $E = 0$, the partial derivative vanishes. However, this occurs at the boundary, so at this point we still do not connect comparable points.

The resulting family of value curves is shown in Figure 67. From the figure, we see that the curves are negatively sloped at each point, and the curves do not cross.

$$f(Q,E) = \frac{-5(1-Q) + 3(1-Q)E + 2(1-Q)^2 - E + E^2}{(E-Q+1)(E-Q-1)} = \gamma \qquad 5.19$$

This cost function arises from analysis of binary classifiers. Figure 68 shows the resulting family of value curves. Several of these curves converge at the points $(1,0)$ and $(0,1)$. However, since this point is at the boundary, the curves do not cross.

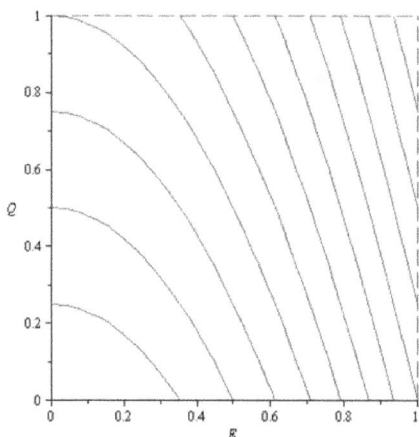

Figure 67: Family of value curves resulting from 5.16.

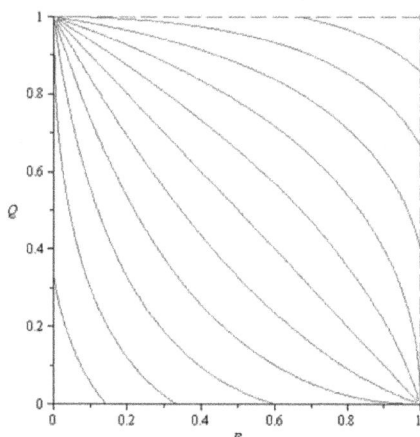

Figure 68: Family of value curves resulting from 5.19.

5.2 Pathological Tradeoff Functions

A tradeoff function that does not meet the criteria from 5.15 is pathological. These tradeoff functions connect points that are Pareto comparable. This leads to a contradiction as the tradeoff function values these points as equivalent, while the Pareto analysis finds one point superior to the other.

Pathological tradeoff functions indicate the presence of a hidden variable. Suppose we have two variables x and y and have identified a few points that have a tradeoff function

$$x^2 - 2y^2 = 0 \qquad 5.20$$

leading to the value curves

$$f(x,y) = x^2 - 2y^2 = \gamma \qquad\qquad 5.21$$

The partial derivatives are

$$\frac{\partial f(x,y)}{\partial x} = 2x \qquad\qquad 5.22$$

$$\frac{\partial f(x,y)}{\partial y} = -4y \qquad\qquad 5.23$$

The partial derivative with respect to y does not meet the criterion from 5.15. This is a pathological tradeoff function and the value curves are positively sloped as shown in 5.21. The points connected by equal value in this figure are not compatible with the concept of Pareto-comparison.

The tradeoff function is meant to compare two points. Extending this to a value curve is proper when the tradeoff function is rational. In this case, the tradeoff function is pathological and should not be extended to a value function.

Pathological tradeoff functions indicate the presence of a hidden variable. Suppose the true tradeoff function is given by

$$f(x,y) = x^2 + y^2 + z^2 \qquad\qquad 5.24$$

The value curves are

$$x^2 + y^2 + z^2 = \gamma \qquad\qquad 5.25$$

The partial derivatives are

$$\frac{\partial f(x,y)}{\partial x} = 2x \qquad\qquad 5.26$$

$$\frac{\partial f(x,y)}{\partial y} = 2y \qquad\qquad 5.27$$

$$\frac{\partial f(x,y)}{\partial z} = 2z \qquad\qquad 5.28$$

This tradeoff function meets the criteria of 5.15 and is shown in Figure 70. A three-dimensional tradeoff function such as this can lead to incorrectly identifying the tradeoff function from 5.20. The projection of this function onto the x-y plane is quarter circles as shown in Figure 72.

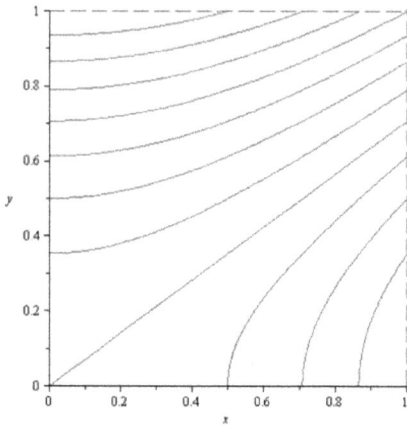

Figure 69: Family of value curves resulting from 5.21.

Figure 70: Family of value curves resulting from 5.25.

All points within the circles in Figure 71 may be equivalent because they may all belong to the same level curve of the true three-dimensional value curve. However, if we identify the points in Figure 72 as equivalent, and if we believe there are only two variables x and y, we may incorrectly conclude that these Pareto-comparable points are actually connected by positively sloped lines. However, when the full three-dimensional nature of the tradeoff is identified, it becomes clear that there is no contradiction at all. In identifying only a few points and drawing conclusions from those, we were led to value curves that do not make sense. Generally, when we find value curves that appear pathological, we should consider the potential presence of hidden variables.

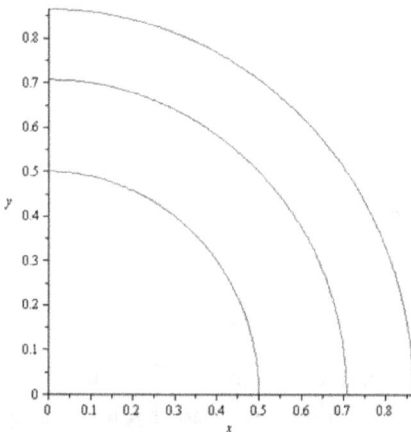

Figure 71: Family of value curves resulting from projecting 5.25 onto the x-y plane.

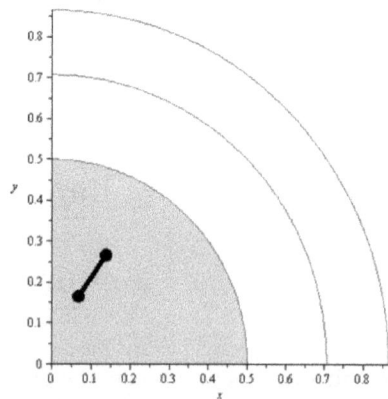

Figure 72: Two equivalent points may appear to be connected with a positively sloped line because a hidden variable is unaccounted.

5.3 Relative Tradeoff Functions

The previous sections have loosely discussed tradeoff functions and value curves. In this section we formalize these definitions.

Suppose we have a multiobjective optimization problem on n-dimensions. A performance point is an achievable point in the n-dimensional vector space and is represented as the vector $\vec{x}$.

A tradeoff function is generally a vector valued function mapping two n-dimensional performance points to a m-dimensional vector valued function $f : \mathbb{R}^n \times \mathbb{R}^n \to \mathbb{R}^m$.

A simple example of a tradeoff function is taking the components of each of the two points and examining the differences in each component value. This is a typical tradeoff analysis where we look at how much we gain in some components in relation to how much we lose in other components.

Many tradeoff functions are single-valued functions. In this case $f : \mathbb{R}^n \times \mathbb{R}^n \to \mathbb{R}$. Single-valued tradeoff functions are often used to quantify how much of one variable we are willing to give up to receive benefit in another variable.

These tradeoff functions may be expressed in terms of differences between components. These tradeoff functions take the form $f(\vec{x}_1 - \vec{x}_2)$. Here, only the difference in the components are present in the function.

Tradeoff functions are usually intended to work with any two performance points. If we set $\vec{x}_2 = \vec{0}$, the above tradeoff becomes $f(\vec{x}_1)$. The tradeoff functions we have examined thus far take this form.

Value functions arise from tradeoff functions of the form $f(\vec{x}_1)$ by computing the curves of constant value $f(\vec{x}_1) = \gamma$. The points on this curve are considered equally valued relative to each other. This means that we are indifferent to two points where $f(\vec{x}_1) = \gamma$ and $f(\vec{x}_2) = \gamma$. Both of these points have the same level of utility from the perspective of multiobjective optimization.

The value γ computed from a value function indicates the desirability for the point. If we have two points, $\vec{x}_1$ and $\vec{x}_2$, if $f(\vec{x}_1) > f(\vec{x}_2)$ then $\vec{x}_1$ is more desirable than $\vec{x}_2$.

The value function must be transitive so if $f(\vec{x}_1) > f(\vec{x}_2)$ and $f(\vec{x}_2) > f(\vec{x}_3)$ then $f(\vec{x}_1) > f(\vec{x}_3)$. The transitive property is useful in rank ordering the frontier as described in the next section.

5.4 Rank Ordering the Frontier

Tradeoff and value function may be used to rank order the points on the Pareto frontier. The different types of tradeoff functions available lead to different methods for rank ordering the points. The subsections below provide some methods that may be employed to rank order the performance points for multiobjective optimization problems.

5.4-a VALUE FUNCTIONS

Value functions may be used to rank order the points on the Pareto frontier. Once a value function and the frontier are identified, we simply put the vectors for the points on the frontier into the value function to compute the value for each of these points. Higher value indicates more desirable points.

Value functions are perhaps the easiest method for rank ordering the points on the frontier. Here, we have identified a single-values function $f(\vec{x})$ where the value of the function at a point $\vec{x}$ represents the fitness for the point.

We simply compute $f(\vec{x})$ for every point on the frontier, and rank order the results. The rank ordering of the value function corresponds to the rank ordering of the points.

5.4-b METRIC TRADEOFF FUNCTIONS

Let $\vec{x}_1$, $\vec{x}_2$, and $\vec{x}_3$ be three points in the optimization space (the vector space created by treating each dimension as the component of a vector). Let $f(\vec{x}, \vec{y})$ be a single-valued tradeoff function relating two performance vectors.

The value $f(\vec{x}, \vec{y})$ is a single numerical quantity representing the value difference between these performance points. This might be viewed as a 'gain' or 'loss' in moving between the performance points $\vec{x}$ and $\vec{y}$. If this is the case, we would expect

$$f(\vec{x}, \vec{y}) = -f(\vec{y}, \vec{x}) \qquad\qquad 5.29$$

This is to say that if we gain a value $f(\vec{x}, \vec{y})$ moving from $\vec{x}$ to $\vec{y}$, then we expect to lose a similar value if we move from $\vec{y}$ to $\vec{x}$. Similarly, we expect $f(\vec{x}, \vec{x}) = 0$ as there is no tradeoff in moving from a point back to itself. Moreover, we expect that $f(\vec{x}, \vec{y}) \neq 0$ when $\vec{x} \neq \vec{y}$. This means that there is always some tradeoff when we move between two different points.

Examine the absolute value of the tradeoff function $|f(\vec{x}, \vec{y})|$. Let $a = |f(\vec{x}_1, \vec{x}_2)|$ and $b = |f(\vec{x}_2, \vec{x}_3)|$. We expect the tradeoff $c = |f(\vec{x}_1, \vec{x}_3)|$ is less than or equal to the sum $a + b$. For example if we recive a benefit of a moving from point 1 to point 2, and a benefit b from moving from point 2 to point 3, then we expect the benefit we get moving from point 1 to point 3 to be no greater than $a + b$.

This is an intuitive notion that is not necessarily true. However, when a tradeoff function satisfies these criteria, the absolute value of the tradeoff function forms a metric space.

Let $f(\vec{x}, \vec{y})$ be a tradeoff function satisfting

$$f(\vec{x}, \vec{y}) = 0 \; iff \; \vec{x} = \vec{y} \qquad 5.30$$

$$|f(\vec{x}, \vec{z})| \leq |f(\vec{x}, \vec{y})| + |f(\vec{y}, \vec{z})| \qquad 5.31$$

$$f(\vec{x}, \vec{y}) = -f(\vec{y}, \vec{x}) \qquad 5.32$$

for all vectors $\vec{x}$, $\vec{y}$, and $\vec{z}$ in the optimization space. Such a tradeoff function is called a metric tradeoff.

Functions satisfying the criteria 5.30 and 5.31 are distance functions. These distance functions form the basis for metric spaces.

Metric tradeoff functions may be used to rank order the frontier. We begin with any point on the frontier. Use the metric to compute the distance from this point to every other point on the frontier. If all other frontier points have negative tradeoff, the point we started with is the highest ranked point. Otherwise, eliminate all points that compute a negative tradeoff. Repeat the process with the positive tradeoff points until we eliminate all others. This is the highest ranked point.

If we eliminate this point from the set and perform the process again, the point identified is the second highest ranked point. We can iterate this to identify a rank ordered list of points among the frontier.

5.5 Partial Tradeoff Functions

Tradeoff functions need not use every available dimension under optimization. Tradeoff functions that only utilize a subset of the available dimensions are called partial tradeoff functions.

Partial tradeoff functions cannot be used to rank order the frontier. The frontier points depend on all available dimensions. Partial tradeoff functions do not provide any information on how the excluded dimensions relate.

Partial tradeoff functions that are single-valued may be used to create partial value functions. To examine these functions, suppose we have a n-dimensional optimization space and a partial tradeoff function $f(\vec{x})$ that relates k of these dimensions. We can use such a function to create the partial value function $f(\vec{x}) = \gamma$.

A rational partial tradeoff function must satisfy

$$\alpha_i = \frac{\partial f(\vec{x})}{\partial x_i} > 0 \quad i = 1 \dots k \qquad\qquad 5.33$$

for each of the dimensions appearing in the partial tradeoff function.

For a given performance point $\vec{x}$, we can compute the partial value for the k-dimensional subvector. This yields a single value for the set of k components of the performance point. We may use this value to replace the k components with this single value.

Generally, this process maps a n-dimensional performance point to a point in a $(n - k + 1)$-dimensional space. For a traditional tradeoff function, $k = n$, so we end up mapping a n-dimensional performance point to a point in one-dimension. Since there is only one dimension, we can use this to optimize the points with traditional methods. This is why tradeoff functions allow us to rank order the frontier.

As an example of a partial tradeoff function, suppose we have a three-dimensional multiobjective optimization problem. Let the variables x, y, and z represent the different dimensions of the optimization space, where each variable is on the range $[0,1]$. Suppose we have the partial tradeoff function

$$f(x, y) = x^2 + y^2 \qquad\qquad 5.34$$

This leads to the partial value function

$$x^2 + y^2 = \gamma \qquad\qquad 5.35$$

The value function may be used to transform a point in the three-dimensional optimization space (x, y, z) to a point in the two-dimensional space $(\gamma, z) = (x^2 + y^2, z)$. In this sense we have transformed the three-dimensional optimization problem to a two-dimensional optimization problem.

6 Identifying Tradeoff Functions

6.1 Inherent Tradeoff Functions

Many tradeoff functions arise naturally from the characteristics of the optimization problem. These inherit tradeoff functions are easily identified and may be used to rank order the frontier to determine an optimal operating point.

As an example, suppose we have an automated device that examines parts for assembly into a product. The device examines a part and determines if the part is fit for production or should be rejected.

There are four possible outcomes for examining each part: the device accepts a part that is fit, the device rejects a part this is actually fit, the device accepts a part that is actually unfit, and the device rejects a part that is unfit. These four possibilities are respectively called true positives, false positives, false negatives, and true negatives.

A device such as this is called a binary classifier. The performance of a binary classifier is characterized with only two numbers: the true positive rate (φ), and the true negative rate (ω). These variables are on the range [0,1], and an optimal classifier has values $\varphi = \omega = 1$.

Further, suppose the device has several different modes of operation. Each mode of operation has different values for φ and ω, and the goal is to maximize both.

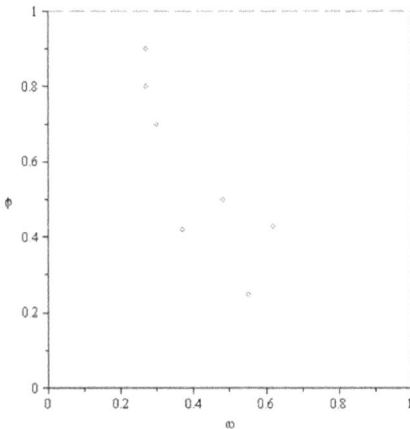

Figure 73: Example binary classifier.

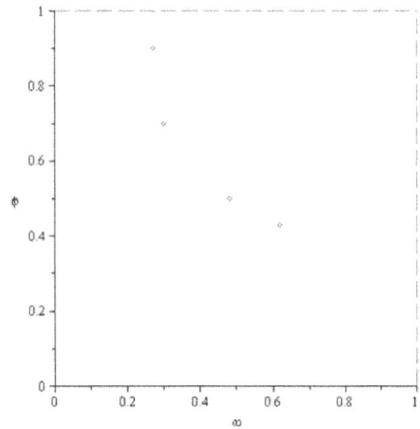

Figure 74: Classifier frontier.

Figure 73 shows the performance points for our device, while Figure 74 provides the Pareto frontier. These points represent the potential operating points for the device.

A natural tradeoff function arises when we consider the costs associated with the performance of the classifier. For example, suppose that each unit constructed correctly can be sold for a profit of $10. Each part that is discarded costs $1, and each unit assembled with an unfit part costs $3 in warranty repairs. Finally, a part that is incorrectly identified costs $11: $1 for rejecting the part, and $10 in lost profits. These values may be used to construct the expected total profit. Each outcome of the classifier is associated with a profit or cost:

Classifier Result	Variable	Profit
True Positive	φ	10
False Positive	$1 - \omega$	-3
False Negative	$1 - \varphi$	-11
True Negative	ω	-1

The expected return using this classifier is

$$10\varphi - 3(1 - \omega) - 11(1 - \varphi) - \omega \qquad \text{6.1}$$

$$21\varphi - 14 + 2\omega \qquad \text{6.2}$$

The expected return is a natural candidate for the tradeoff function. The constant value (-14) does not affect the tradeoff and may be ignored. The tradeoff function is

$$f(\varphi, \omega) = 21\varphi + 2\omega \qquad \text{6.3}$$

The partial derivatives are

$$\frac{\partial f(\varphi, \omega)}{\partial \varphi} = 21 \qquad \text{6.4}$$

$$\frac{\partial f(\varphi, \omega)}{\partial \omega} = 3 \qquad \text{6.5}$$

The partial derivatives meet the criteria from 5.15, so the tradeoff function is rational. Value curves may be generated from

$$21\varphi + 2\omega = \gamma \qquad \text{6.6}$$

These are negatively sloped lines in the ω-φ plane. Figure 75 shows the frontier points with their associated value curves.

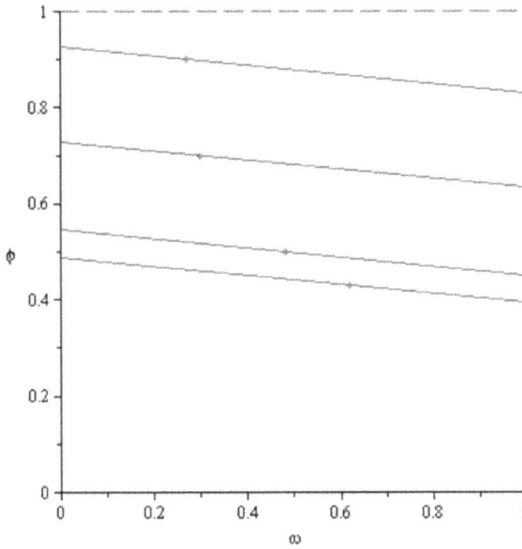

Figure 75: Value curves for the classifier frontier.

The point $(\omega, \varphi) = (.27, .90)$ lies on the highest value curve. The φ component dominates the tradeoff function because the coefficient of this term is much larger than the ω term.

This simple example illustrates how tradeoff functions may naturally arise from the optimization problem under consideration. These cases often reduce the multiobjective optimization problem to a single objective optimization.

6.2 Pareto Condition

When no obvious tradeoff/value function exists, we may choose to construct a value function through other means. Rational value functions only connects points that are Pareto incomparable. This leads to the Pareto condition. In this section we express this condition mathematically.

We begin by computing this value function condition in two and three dimensions. We then generalize the process for three dimensions to an arbitrary number of dimensions.

6.2-a PARETO CONDITION IN TWO DIMENSIONS

In two dimensions, the condition that a value function only connect incomparable points means that the value function must be negatively sloped at every point. As we move to higher dimensions, the condition is not as succinct. To aid with extending this concepts to higher dimensions, we express the condition in two dimensions more generally.

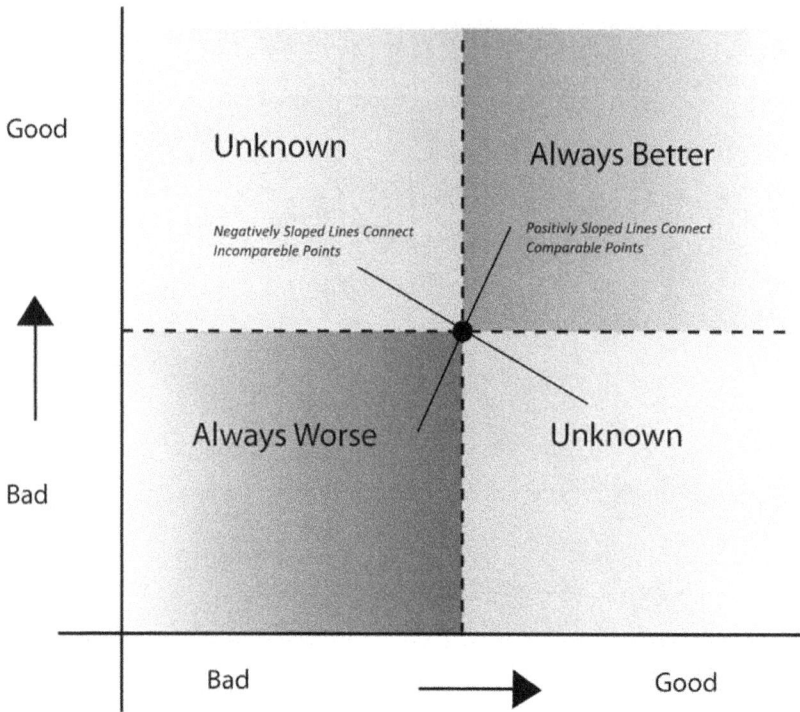

Figure 76: Rational value curves connect incomparable points. Value curves represent points that are considered to have the same 'value'. Value curve the connect Pareto-comparable points lead to a contradiction. One point is superior in the Pareto sense, but the points are equivalent from the value curve.

Let $f(x, y)$ be a rational tradeoff function. The total derivative is

$$df = \frac{\partial f}{\partial x} dx + \frac{\partial f}{\partial y} dy$$ 6.7

The partial derivatives are related to the gradient:

$$\vec{\nabla} f = \frac{\partial f}{\partial x} \hat{x} + \frac{\partial f}{\partial y} \hat{y}$$ 6.8

In two dimensions, the gradient of a function is a vector that points in the direction perpendicular to the tangent of the function. In general, there are two possible vectors tangent to a line. Both vectors are perpendicular to the line, but point in opposite directions. An example of the two vectors perpendicular to the tangent is shown in Figure 77.

Examine the vector pointing in the extreme Pareto-comparable direction $\vec{P} = [1,1]$. For rational value curves, the angle between the gradient and $\vec{P}$ must be either less than 90° or greater than 270° as shown in Figure 78.

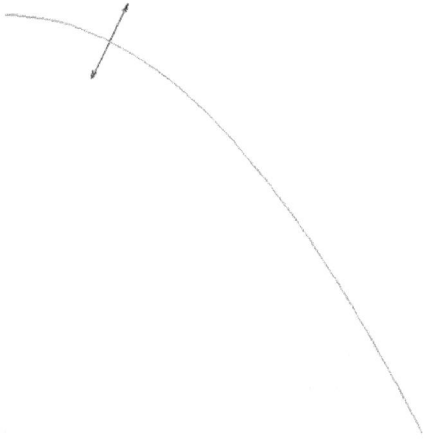

Figure 77: There are two possible directions for a vector normal to a curve.

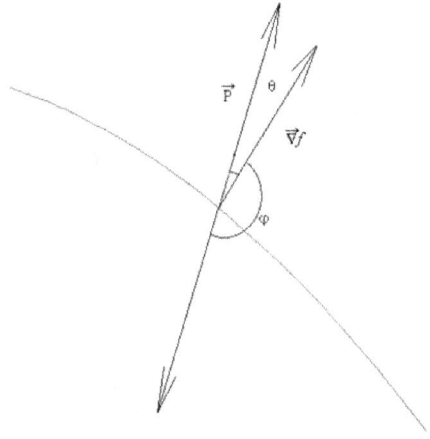

Figure 78: For rational value curves, the angle between the gradient vector and the Pareto direction must be less than 90° or greater than 270°.

The dot product between the gradient vector and $\vec{P}$ may be written in terms of the cosine of the angle between the vectors. The dot product is

$$\vec{\nabla}f \cdot \vec{P} = \frac{\partial f}{\partial x} + \frac{\partial f}{\partial y} = |\vec{\nabla}f||\vec{P}| \cos\theta \qquad \text{6.9}$$

Figure 79 shows the values of the cosine where the value curve is rational. Again, there are separate regions (one positive, the other negative) because there are two perpendicular vectors. However, if we square the result, we only use one range of values as shows in Figure 80. For the square of the cosine, the condition is simply that the square of the cosine is greater than ½. Thus,

$$\cos^2\theta = \frac{\left(\frac{\partial f}{\partial x} + \frac{\partial f}{\partial y}\right)^2}{|\vec{\nabla}f|^2 |\vec{P}|^2} > \frac{1}{2} \qquad \text{6.10}$$

We know that

$$|\vec{\nabla}f|^2 = \left(\frac{\partial f}{\partial x}\right)^2 + \left(\frac{\partial f}{\partial y}\right)^2 \qquad \text{6.11}$$

and

$$|\vec{P}|^2 = 1^2 + 1^2 = 2 \qquad \text{6.12}$$

Substituting into the expression above,

$$\frac{\left(\frac{\partial f}{\partial x}+\frac{\partial f}{\partial y}\right)^2}{\left|\vec{\nabla}f\right|^2\left|\vec{P}\right|^2} > \frac{1}{2} \qquad 6.13$$

$$\frac{\left(\frac{\partial f}{\partial x}+\frac{\partial f}{\partial y}\right)^2}{\left[\left(\frac{\partial f}{\partial x}\right)^2+\left(\frac{\partial f}{\partial y}\right)^2\right](2)} > \frac{1}{2} \qquad 6.14$$

$$\frac{\left(\frac{\partial f}{\partial x}\right)^2+2\left(\frac{\partial f}{\partial x}\right)\left(\frac{\partial f}{\partial y}\right)+\left(\frac{\partial f}{\partial y}\right)^2}{\left[\left(\frac{\partial f}{\partial x}\right)^2+\left(\frac{\partial f}{\partial y}\right)^2\right]} > 1 \qquad 6.15$$

$$1+\frac{2\left(\frac{\partial f}{\partial x}\right)\left(\frac{\partial f}{\partial y}\right)}{\left[\left(\frac{\partial f}{\partial x}\right)^2+\left(\frac{\partial f}{\partial y}\right)^2\right]} > 1 \qquad 6.16$$

$$\left(\frac{\partial f}{\partial x}\right)\left(\frac{\partial f}{\partial y}\right) > 0 \qquad 6.17$$

The last expression is the Pareto condition in two dimensions. A rational two dimensional value curve must meet the condition of 6.17 at every point in the operational region.

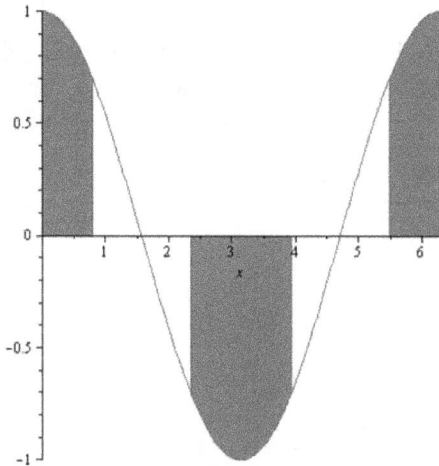

Figure 79: Values of the cosine function where the angle between the gradient and the Pareto vector yield a rational value curve.

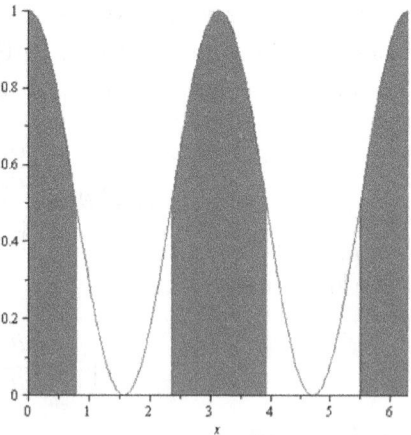

Figure 80: Values of the square of the cosine function where the angle between the gradient and the Pareto vector yield a rational value curve.

6.2-b   PARETO CONDITION IN THREE DIMENSIONS

Condition 6.17 may be extended to three dimensions with minor modifications. Again, the gradient of the value function is compared to the three dimensional Pareto vector. In three dimensions the Pareto vector is $\vec{P} = [1,1,1]$. We need to find the range of the cosine (or cosine squared) that is allowable.

Figure 81 shows the Pareto vector in three dimensions and the allowable angles for rational value curves. If we compute the dot product of the Pareto vector with a unit vector along one of the axes we find

$$\vec{P} \cdot \hat{x} = |\vec{P}||\hat{x}| \cos \theta \qquad\qquad 6.18$$

The dot product is

$$\vec{P} \cdot \hat{x} = 1 \qquad\qquad 6.19$$

The magnitude of the Pareto vector is

$$|\vec{P}| = \sqrt{1^2 + 1^2 + 1^2} = \sqrt{3} \qquad\qquad 6.20$$

Thus,

$$\cos \theta = \frac{1}{\sqrt{3}} \qquad\qquad 6.21$$

The dot product between the Pareto vector and the gradient of the value function is

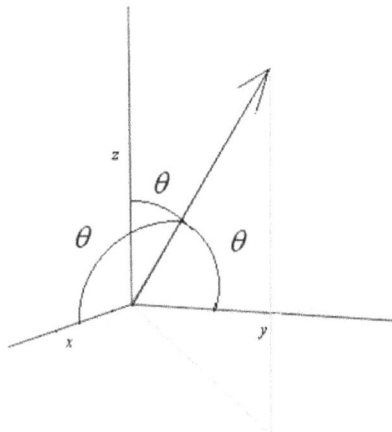

Figure 81: The three dimensional Pareto vector and the allowable angles for rational value curves.

$$\vec{\nabla} f \cdot \vec{P} = \frac{\partial f}{\partial x} + \frac{\partial f}{\partial y} + \frac{\partial f}{\partial z} = |\vec{\nabla} f||\vec{P}| \cos \theta \qquad 6.22$$

Thus,

$$\cos^2 \theta = \frac{\left(\frac{\partial f}{\partial x} + \frac{\partial f}{\partial y} + \frac{\partial f}{\partial z}\right)^2}{|\vec{\nabla} f|^2 |\vec{P}|^2} > \frac{1}{3} \qquad 6.23$$

$$\left(\frac{\partial f}{\partial x} + \frac{\partial f}{\partial y} + \frac{\partial f}{\partial z}\right)^2 > \left(\frac{\partial f}{\partial x}\right)^2 + \left(\frac{\partial f}{\partial y}\right)^2 + \left(\frac{\partial f}{\partial z}\right)^2 \qquad 6.24$$

$$\left(\frac{\partial f}{\partial x}\right)\left(\frac{\partial f}{\partial y}\right) + \left(\frac{\partial f}{\partial x}\right)\left(\frac{\partial f}{\partial z}\right) + \left(\frac{\partial f}{\partial y}\right)\left(\frac{\partial f}{\partial z}\right) > 0 \qquad 6.25$$

The inequality from 6.25 is the Pareto condition in three dimensions.

6.2-c   PARETO CONDITION IN N-DIMENSIONS

Finally, we extend the previous results to an arbitrary number of dimensions. For the n-dimensional case, the Pareto vector is $\vec{P} = [1, \dots, 1]$. The allowable values for the cosine may be found by taking the dot product of the Pareto vector with a unit vector along one of the axes:

$$\vec{P} \cdot \hat{x} = |\vec{P}||\hat{x}| \cos \theta = 1 \qquad 6.26$$

where

$$|\vec{P}|^2 = n \qquad 6.27$$

Thus,

$$\cos^2 \theta > \frac{1}{n} \qquad 6.28$$

Taking the dot product between the Pareto vector and the gradient,

$$\vec{\nabla} f \cdot \vec{P} = \sum_{i=1}^{n} \frac{\partial f}{\partial x_i} = |\vec{\nabla} f||\vec{P}| \cos \theta \qquad 6.29$$

Solving for the cosine and squaring,

$$\cos^2 \theta = \frac{\left(\sum_{i=1}^{n} \frac{\partial f}{\partial x_i}\right)^2}{|\vec{\nabla} f|^2 |\vec{P}|^2} > \frac{1}{n} \qquad 6.30$$

$$\left(\sum_{i=1}^{n} \frac{\partial f}{\partial x_i}\right)^2 > \sum_{i=1}^{n} \left(\frac{\partial f}{\partial x_i}\right)^2 \qquad 6.31$$

$$\sum_{i \neq j}^{n} \frac{\partial f}{\partial x_i} \frac{\partial f}{\partial x_j} > 0 \qquad 6.32$$

The final sum is over all values of i and j where $i \neq j$. The inequality of 6.32 is the Pareto condition in n-dimensions.

6.3 Lagrange Multipliers

In some cases there is no obvious tradeoff function. We may have a set of operational points we can use to find a frontier, but no tradeoff function to use to rank order the points on the frontier.

One approach to identifying a tradeoff function in this case is to identify pairs of incomparable points that the user feels are roughly equivalent. If we have a set of pairs, we may use this information to fit a family of curves using least squares and the method of Lagrange multipliers.

Suppose we have an n-dimensional multiobjective optimization problem and have m potential operational points identified. We can examine each point and ask the user if there are any incomparable points that the user feels are equivalent in preference. Not every point may have an equivalent.

Assume the user identifies k such points of points. Let $\vec{x}_{1l}$ and $\vec{x}_{2l}$ represent a pair of such points. Furthermore, let $f(\vec{x})$ be the form for the tradeoff function. This function is parameterized by p constants that we desire to identify by fitting the function to the pairs of equivalent points.

For example, in two dimensions, we may choose a fitness function of the form

$$f(x, y) = \alpha x^2 + \beta xy + \gamma y^2 + \delta x + \varepsilon y + \eta \qquad 6.33$$

Here, the parameters are $\alpha, \beta, \gamma, \delta, \varepsilon$, and η. This is the form of a generalized quadratic in two dimensions. We want to use the pairs of points identified to determine the best values for these parameters. Once we do this, we have a tradeoff function that we can use to rank order the frontier.

Not only do we need to fit the function to the pairs of equivalent points identified, we need the tradeoff function to obey the Pareto constraint

$$\left(\frac{\partial f}{\partial x}\right)\left(\frac{\partial f}{\partial y}\right) > 0 \qquad 6.34$$

To formalize the problem, let $\vec{\rho}$ be a vector of p parameter values. Let the fitness function be represented as $f(\vec{x};\vec{\rho})$ were $\vec{x}$ is a n-dimensional vector to an operating point, and $\vec{\rho}$ is a p-dimensional vector of parameters. Suppose we have k constraints each equating a pair of operational points:

$$f(\vec{x}_{1j}) = f(\vec{x}_{2j}) \quad j = 1 \dots k \tag{6.35}$$

and the rationality constraint

$$\left(\frac{\partial f}{\partial x}\right)\left(\frac{\partial f}{\partial y}\right) > 0 \tag{6.36}$$

The least squares method leads to minimizing

$$\chi^2 = \sum_{j=1}^{k}\left[f(\vec{x}_{1j}) - f(\vec{x}_{2j})\right]^2 \tag{6.37}$$

subject to the constraint

$$\left(\frac{\partial f}{\partial x}\right)\left(\frac{\partial f}{\partial y}\right) > 0 \tag{6.38}$$

The constraint leads to a constraint on the values of the parameters. The Pareto constraint must be met at every point in the operational space, not just the points that are pair-equivalent.

In most cases, minimizing χ^2 subject to the constraint from 6.38 is computed using numerical methods. In this case, we may enforce 6.38 by computing the partial derivative of the value curve at several points in the operational region, then verifying that the condition is met. The techniques from Chapter 2 may be used to minimize χ^2 while varying the parameters for the value curve. For a given set of values for the parameters, we confirm that the constraint is met.

The parameter constraint resulting from 6.38 depends on the form of the tradeoff function used. This analysis is continued in the next sections assuming different forms for the tradeoff function.

In each of these sections, we assume that each component of $\vec{x}$ is on the range [0,1]. This assumption allows us to determine the form of the parameter constraints from 6.38. Many multiobjective optimization problems result in dependent variables on the range [0,1].

Generally, when the dependent variables are not on this range, there is some finite maximum and minimum value for the range of the variables. In this case, we may scale the operational point to the range [0,1]. The only case where this does not work is when a dependent variable is allowed to have an infinite value.

6.3-a   LINEAR TRADEOFF

Assume the dependent variables are on the range [0,1], and the tradeoff function has the form

$$f(\vec{x}) = \sum_{i=1}^{n} \rho_i x_i \qquad 6.39$$

In this case $p = n$ because we have one parameter for each dimension. This is a general linear tradeoff function. The constraint 6.38 leads to the conditions

$$\sum_{i \neq j}^{n} \rho_i \rho_j > 0 \qquad 6.40$$

We minimize χ^2 with respect to each of the parameters in $\vec{\rho}$. The result is a set of n equations we simultaneously solve for the n parameters. With this tradeoff function χ^2 is

$$\chi^2 = \sum_{j=1}^{k} [f(\vec{x}_{1j}) - f(\vec{x}_{2j})]^2 \qquad 6.41$$

$$= \sum_{j=1}^{k} \left[\sum_{i=1}^{n} \rho_i [x_{1ji} - x_{2ji}] \right]^2 \qquad 6.42$$

where x_{1ji} means the ith component of the first point from the jth equation. Minimizing χ^2 subject to the constraint from 6.40 creates a system of equations. Next, we take the partial derivatives with respect to each parameter and set them to zero,

$$\frac{\partial \chi^2}{\partial \rho_i} = \sum_{j=1}^{k} 2[f(\vec{x}_{1j}; \vec{\rho}) - f(\vec{x}_{2j}; \vec{\rho})] \left[\frac{\partial f(\vec{x}_{1j}; \vec{\rho})}{\partial \rho_i} - \frac{\partial f(\vec{x}_{2j}; \vec{\rho})}{\partial \rho_i} \right] = 0 \qquad 6.43$$

$$\sum_{j=1}^{k} \sum_{l=1}^{n} \rho_l [x_{1jl} - x_{2jl}][x_{1ji} - x_{2ji}] = 0 \qquad 6.44$$

This is a matrix equation and has the trivial solution $\vec{\rho} = 0$. If all of the parameters are zero, the tradeoff function becomes zero as well. This satisfies 6.44, but does not satisfy the constraint 6.40.

To find a solution to the system meeting the constraint, set the value of the tradeoff function to a non-zero value at some point. This leads to the equation

$$f(\vec{x}) = \sum_{i=1}^{n} \rho_i \bar{x}_i = \omega \qquad \text{6.45}$$

where the bar is intended to indicate that we are choosing some particular values for the components. This adds an additional constraint to the system. We can incorporate this constraint using the method of Lagrange. The method of Lagrange multiplies the constraint equation by an unknown value λ and adds this to χ^2. The new χ^2 is

$$\chi^2 = \sum_{j=1}^{k} [f(\vec{x}_{1j}) - f(\vec{x}_{2j})]^2 + 2\lambda \left[\sum_{i=1}^{n} \rho_i \bar{x}_i - \omega \right] \qquad \text{6.46}$$

Taking the partial derivatives,

$$\frac{\partial \chi^2}{\partial \rho_i} = \sum_{j=1}^{k} 2[f(\vec{x}_{1j}; \vec{\rho}) - f(\vec{x}_{2j}; \vec{\rho})] \left[\frac{\partial f(\vec{x}_{1j}; \vec{\rho})}{\partial \rho_i} - \frac{\partial f(\vec{x}_{2j}; \vec{\rho})}{\partial \rho_i} \right] + \lambda \bar{x}_i = 0 \qquad \text{6.47}$$

$$\sum_{j=1}^{k} \sum_{l=1}^{n} \rho_l [x_{1jl} - x_{2jl}][x_{1ji} - x_{2ji}] + \lambda \bar{x}_i = 0 \qquad \text{6.48}$$

We also minimize with respect to λ

$$\frac{\partial \chi^2}{\partial \lambda} = \sum_{i=1}^{n} \rho_i \bar{x}_i - \omega = 0 \qquad \text{6.49}$$

This leads to the system of equations

$$\begin{bmatrix} \mathcal{M}_{11} & \mathcal{M}_{12} & \cdots & \mathcal{M}_{1n} & \bar{x}_1 \\ \mathcal{M}_{21} & \ddots & & & \bar{x}_2 \\ \vdots & & \ddots & & \vdots \\ \mathcal{M}_{n1} & \mathcal{M}_{n2} & \cdots & \mathcal{M}_{nn} & \bar{x}_n \\ \bar{x}_1 & \bar{x}_2 & \cdots & \bar{x}_n & 0 \end{bmatrix} \begin{bmatrix} \rho_1 \\ \rho_2 \\ \vdots \\ \rho_n \\ \lambda \end{bmatrix} = \begin{bmatrix} 0 \\ 0 \\ \vdots \\ 0 \\ \omega \end{bmatrix} \qquad \text{6.50}$$

where

$$\mathcal{M}_{ij} = \sum_{m=1}^{k} [x_{1mj} - x_{2mj}][x_{1mi} - x_{2mi}] \qquad \text{6.51}$$

This matrix is symmetric so that

$$\mathcal{M}_{ij} = \mathcal{M}_{ji} \qquad \text{6.52}$$

Alternatively, we can express the matrix as

$$M_{ij} = \begin{pmatrix} \sum_{m=1}^{k} (\Delta x_m)^2 & \sum_{m=1}^{k} \Delta x_m \Delta y_m \\ \sum_{m=1}^{k} \Delta x_m \Delta y_m & \sum_{m=1}^{k} (\Delta y_m)^2 \end{pmatrix} \qquad 6.53$$

The diagonal elements must be positive because these are the sum of the squares of number. The off diagonal elements must be negative because each term is negative. These terms are negative because the equivalent points selected by the user must be Pareto incomparable points. For Pareto incomparable points in two dimensions, either Δx or Δy must be negative, but never both.

This can be solved by inverting the matrix on the left:

$$\begin{bmatrix} \rho_1 \\ \rho_2 \\ \vdots \\ \rho_n \\ \lambda \end{bmatrix} = \begin{bmatrix} M_{11} & M_{12} & \cdots & M_{1n} & \bar{x}_1 \\ M_{21} & \ddots & & & \bar{x}_2 \\ \vdots & & \ddots & & \vdots \\ M_{n1} & M_{n2} & \cdots & M_{nn} & \bar{x}_n \\ \bar{x}_1 & \bar{x}_2 & \cdots & \bar{x}_n & 0 \end{bmatrix}^{-1} \begin{bmatrix} 0 \\ 0 \\ \vdots \\ 0 \\ \omega \end{bmatrix} \qquad 6.54$$

As a simple example, suppose we have a two dimensional optimization problem with the linear tradeoff

$$f(x, y) = \rho_1 x + \rho_2 y \qquad 6.55$$

Suppose we determine that the points (1,0) and (0,1) are equivalent. Further, suppose we set the value of the tradeoff at the point (1,1) to ω.

This system is easily solved directly. The equivalent points lead to the equation

$$f(1,0) = \rho_1 = f(0,1) = \rho_2 \qquad 6.56$$

so

$$\rho_1 = \rho_2 \qquad 6.57$$

Using the condition at (1,1),

$$f(1,1) = \rho_1 + \rho_2 = 2\rho_1 = \omega \qquad 6.58$$

The solution is

$$\rho_1 = \rho_2 = \frac{\omega}{2} \qquad 6.59$$

Let's solve this system using 6.54. First, compute the matrix M:

$$M_{11} = \sum_{m=1}^{1} [x_{1m1} - x_{2m1}][x_{1m1} - x_{2m1}] \qquad 6.60$$

$$\mathcal{M}_{11} = [x_{111} - x_{211}][x_{111} - x_{211}] \tag{6.61}$$

$$\mathcal{M}_{11} = [1 - 0][1 - 0] = 1 \tag{6.62}$$

Similarly,

$$\mathcal{M}_{12} = [0 - 1][1 - 0] = -1 \tag{6.63}$$

$$\mathcal{M}_{21} = [0 - 1][1 - 0] = -1 \tag{6.64}$$

$$\mathcal{M}_{22} = [0 - 1][1 - 0] = 1 \tag{6.65}$$

Thus,

$$\begin{bmatrix} \rho_1 \\ \rho_2 \\ \lambda \end{bmatrix} = \begin{bmatrix} 1 & -1 & 1 \\ -1 & 1 & 1 \\ 1 & 1 & 0 \end{bmatrix}^{-1} \begin{bmatrix} 0 \\ 0 \\ \omega \end{bmatrix} \tag{6.66}$$

$$\begin{bmatrix} \rho_1 \\ \rho_2 \\ \lambda \end{bmatrix} = \frac{1}{4} \begin{bmatrix} 1 & -1 & 2 \\ -1 & 1 & 2 \\ 2 & 2 & 0 \end{bmatrix} \begin{bmatrix} 0 \\ 0 \\ \omega \end{bmatrix} \tag{6.67}$$

$$\begin{bmatrix} \rho_1 \\ \rho_2 \\ \lambda \end{bmatrix} = \begin{bmatrix} \omega/2 \\ \omega/2 \\ 0 \end{bmatrix} \tag{6.68}$$

Again we find

$$\rho_1 = \rho_2 = \frac{\omega}{2} \tag{6.69}$$

This is the same solution we found by directly solving the equations.

It may be troubling that the parameters depend on the value ω. Does our choice of this value effect the results? Keep in mind, the point of the tradeoff function is to connect equivalent points. Can the choice of ω change the points we are connecting?

In this case, the tradeoff function is a set of straight lines. The slope of the lines is what determines which points are connected. Thus, what really matters is not the actual value of the parameters, but the ratio of the parameters. This is because we are dealing with a family of lines, not a single line. Thus the tradeoff functions

$$f(x, y) = x + y \tag{6.70}$$

$$f(x, y) = 2x + 2y \tag{6.71}$$

$$f(x, y) = 100x + 100y \tag{6.72}$$

all generate the same family of curves. The slopes of the lines in each family depends only on the ratio of the parameters. In our case, the ratio of the parameters is

$$\frac{\rho_1}{\rho_2} = 1 \qquad \qquad 6.73$$

which does not depend on the value of ω we choose.

What about the choice for the point $(1,1)$. We simply choose this point and set the value of the tradeoff there to the value ω. We already saw that the value ω does not matter, but what about the choice of the point $(1,1)$? If we use the arbitrary point (x, y) instead we have

$$\begin{bmatrix} \rho_1 \\ \rho_2 \\ \lambda \end{bmatrix} = \begin{bmatrix} 1 & -1 & x \\ -1 & 1 & y \\ x & y & 0 \end{bmatrix}^{-1} \begin{bmatrix} 0 \\ 0 \\ \omega \end{bmatrix} \qquad \qquad 6.74$$

$$\begin{bmatrix} \rho_1 \\ \rho_2 \\ \lambda \end{bmatrix} = - \begin{bmatrix} \dfrac{y^2}{(x+y)^2} & -\dfrac{xy}{(x+y)^2} & \dfrac{1}{x+y} \\ \dfrac{xy}{(x+y)^2} & \dfrac{x^2}{(x+y)^2} & \dfrac{1}{x+y} \\ \dfrac{1}{x+y} & \dfrac{1}{x+y} & 0 \end{bmatrix} \begin{bmatrix} 0 \\ 0 \\ \omega \end{bmatrix} \qquad \qquad 6.75$$

$$\begin{bmatrix} \rho_1 \\ \rho_2 \\ \lambda \end{bmatrix} = \begin{bmatrix} \dfrac{\omega}{x+y} \\ \dfrac{\omega}{x+y} \\ 0 \end{bmatrix} \qquad \qquad 6.76$$

Again, the ratio of the parameters is

$$\frac{\rho_1}{\rho_2} = 1 \qquad \qquad 6.77$$

No matter which point we choose or what value we assigned to the tradeoff function at that point, we are led to the same family of curves.

This is not the end of the story. Let's look at the problem more generally. For the two dimensional case where we set one pair of points equal, we are led to the equation

$$\begin{bmatrix} \rho_1 \\ \rho_2 \\ \lambda \end{bmatrix} = \begin{bmatrix} \Delta_x^2 & \Delta_x \Delta_y & x \\ \Delta_x \Delta_y & \Delta_y^2 & y \\ x & y & 0 \end{bmatrix}^{-1} \begin{bmatrix} 0 \\ 0 \\ \omega \end{bmatrix} \qquad \qquad 6.78$$

where $\Delta_x = x_{11} - x_{21}$ (the difference in the x values of the two points we are setting equal) and similarly for Δ_y. The solution is

$$\begin{bmatrix} \rho_1 \\ \rho_2 \\ \lambda \end{bmatrix} = \omega \begin{bmatrix} -\dfrac{\Delta_y}{\Delta_x y - \Delta_y x} \\ \dfrac{\Delta_x}{\Delta_x y - \Delta_y x} \\ 0 \end{bmatrix} \qquad 6.79$$

The ratio of the slopes is

$$\frac{\rho_1}{\rho_2} = -\frac{\Delta_y}{\Delta_x} \qquad 6.80$$

Again, this is independent of the values of the point we choose to set the tradeoff to ω. Once we have picked two points to set equal, the choice of a point where we set the tradeoff to ω does not affect the resulting family of curves.

The results thus far apply to the case in which we only have one pair of points the user designates as equivalent. In general there may be many such pairs. In two dimensions, the most general case results in the equation

$$\begin{bmatrix} \rho_1 \\ \rho_2 \\ \lambda \end{bmatrix} = \begin{bmatrix} \mathcal{M}_{11} & \mathcal{M}_{12} & x \\ \mathcal{M}_{21} & \mathcal{M}_{22} & y \\ x & y & 0 \end{bmatrix}^{-1} \begin{bmatrix} 0 \\ 0 \\ \omega \end{bmatrix} \qquad 6.81$$

First, multiply both sides by the constant ω^{-1}

$$\begin{bmatrix} \rho_1/\omega \\ \rho_2/\omega \\ \lambda/\omega \end{bmatrix} = \begin{bmatrix} \mathcal{M}_{11} & \mathcal{M}_{12} & x \\ \mathcal{M}_{21} & \mathcal{M}_{22} & y \\ x & y & 0 \end{bmatrix}^{-1} \begin{bmatrix} 0 \\ 0 \\ 1 \end{bmatrix} \qquad 6.82$$

Once we compute the matrix inverse, the result has the form

$$\begin{bmatrix} \rho_1/\omega \\ \rho_2/\omega \\ \lambda/\omega \end{bmatrix} = \begin{bmatrix} \alpha \\ \beta \\ \gamma \end{bmatrix} \qquad 6.83$$

The values α and β do not depend on ω because the matrix $\mathcal{M}$ does not depend on ω. The parameters are

$$\rho_1 = \alpha\omega \qquad 6.84$$

$$\rho_2 = \beta\omega \qquad 6.85$$

The ratio is

$$\frac{\rho_1}{\rho_2} = \frac{\alpha}{\beta} \qquad 6.86$$

which does not depend on ω. Thus, the ratio of the parameters does not depend on ω even in the most general case.

If we compute the inverse of the matrix in 6.83 we find

$$\begin{bmatrix} \rho_1/\omega \\ \rho_2/\omega \\ \lambda/\omega \end{bmatrix} = \frac{1}{(\mathcal{M}_{22}x^2 - 2\mathcal{M}_{12}xy + \mathcal{M}_{11}y^2)} \begin{bmatrix} y^2 & -xy & \mathcal{M}_{22}x - \mathcal{M}_{12}y \\ -xy & x^2 & \mathcal{M}_{11}y - \mathcal{M}_{12}x \\ x & y & \mathcal{M}_{11}\mathcal{M}_{22} - \mathcal{M}_{12}^2 \end{bmatrix} \begin{bmatrix} 0 \\ 0 \\ 1 \end{bmatrix} \qquad 6.87$$

or

$$\begin{bmatrix} \rho_1/\omega \\ \rho_2/\omega \\ \lambda/\omega \end{bmatrix} = \frac{1}{(\mathcal{M}_{22}x^2 - 2\mathcal{M}_{12}xy + \mathcal{M}_{11}y^2)} \begin{bmatrix} \mathcal{M}_{22}x - \mathcal{M}_{12}y \\ \mathcal{M}_{11}y - \mathcal{M}_{12}x \\ \mathcal{M}_{11}\mathcal{M}_{22} - \mathcal{M}_{12}^2 \end{bmatrix} \qquad 6.88$$

The ratio of the parameters is

$$\frac{\rho_1}{\rho_2} = \frac{\mathcal{M}_{22}x - \mathcal{M}_{12}y}{\mathcal{M}_{11}y - \mathcal{M}_{12}x} \qquad 6.89$$

This ratio does not depend on ω, but does depend on x and y unless the element of $\mathcal{M}$ meet certain conditions. If the ratio is independent of x and y, then we must have

$$\frac{\mathcal{M}_{22}x - \mathcal{M}_{12}y}{\mathcal{M}_{11}y - \mathcal{M}_{12}x} = \alpha \qquad 6.90$$

where α is a constant independent of x and y. Rearranging

$$(\mathcal{M}_{22} + \alpha\mathcal{M}_{12})x = (\mathcal{M}_{12} + \alpha\mathcal{M}_{11})y \qquad 6.91$$

For this to be true for all x and y we must have

$$\mathcal{M}_{22} + \alpha\mathcal{M}_{12} = 0 \qquad 6.92$$

$$\mathcal{M}_{12} + \alpha\mathcal{M}_{11} = 0 \qquad 6.93$$

Solving these,

$$\alpha = -\frac{\mathcal{M}_{22}}{\mathcal{M}_{12}} \qquad 6.94$$

or

$$\mathcal{M}_{12} - \frac{\mathcal{M}_{22}}{\mathcal{M}_{12}}\mathcal{M}_{11} = 0 \qquad 6.95$$

$$\mathcal{M}_{12}^2 = \mathcal{M}_{22}\mathcal{M}_{11} \qquad 6.96$$

If the elements of $\mathcal{M}$ meet this, then the ratio of the parameters is independent of x and y. Otherwise the ratio is dependent on x and y.

For the particular case where we only have identified one pair of equivalent points, this equation is in fact satisfied. However, in general, this equation is not necessarily met. This means that in the general case, the choice of x and y does make a difference in which points are connected by the tradeoff function.

The ratio of the slopes may be found by minimizing and maximizing 6.89. The extremes for this provide the range

$$-\frac{\mathcal{M}_{12}}{\mathcal{M}_{11}} \leq \frac{\rho_1}{\rho_2} \leq -\frac{\mathcal{M}_{22}}{\mathcal{M}_{12}} \qquad\qquad 6.97$$

or

$$-\frac{\mathcal{M}_{22}}{\mathcal{M}_{12}} \leq \frac{\rho_1}{\rho_2} \leq -\frac{\mathcal{M}_{12}}{\mathcal{M}_{12}} \qquad\qquad 6.98$$

depending on which ratio is larger.

As an example, suppose the user identified the points in Table 5. Each pair of points represents points that the user had identified as equivalent. These points are shown in the operation space in Figure 82. In the figure, dashed lines match equivalent pairs of points.

Point 1	Point 2
$(.5, .7)$	$(.3, .8)$
$(.8, .6)$	$(.7, .7)$
$(.25, .4)$	$(.3, .3)$
$(.1, .6)$	$(.25, .55)$

Table 5: Sample points identified as equivalent

. Using these points, we can compute the elements of $\mathcal{M}$:

$$\mathcal{M}_{ij} = \begin{pmatrix} .750 & -.0425 \\ -.0425 & .0325 \end{pmatrix} \qquad\qquad 6.99$$

The range of the ratio is

$$.567 \leq \frac{\rho_1}{\rho_2} \leq .765 \qquad\qquad 6.100$$

Figure 83 shows some sample points in the operation space with lines of maximum and minimum slope. From each sample point, the points bound by the dotted lines through the point could be evaluated as equivalent depending on the choice of the point where we set the value of the tradeoff function to ω.

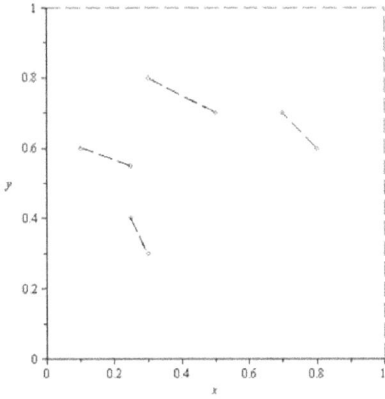

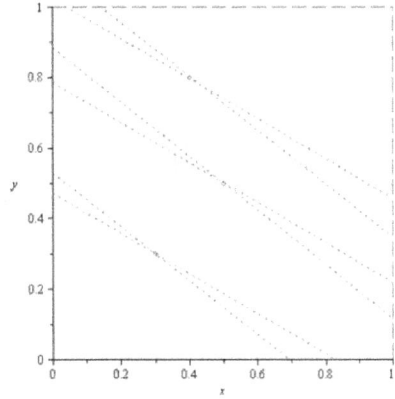

Figure 82: Sample points identified as equivalent.

Figure 83: Points in the operation space showing the maximum and minimum slopes.

This example illustrates that the family of curves determined though this method is not unique. Given a set of points considered equivalent, there is some degree of play in obtaining a linear tradeoff function.

Although the solution is not unique, in many cases the range of the slope ratio is fairly limited. Looking at the points in Figure 83, although there is some play in the slopes, over the range of allowable values of x and y, there are a limited number of points in each region of potential equivalence.

The method of fitting a linear tradeoff to a set of points identified by the decision maker as equivalent provides a useful tradeoff function in many cases of practical interest. In the next section we apply this technique to more general tradeoff functions which contain both linear and quadratic terms.

6.3-b QUADRATIC TRADEOFF

In this section we examine the approach from the previous section applied to a generalized quadratic tradeoff. Again, the dependent variables are on the range $[0,1]$. The tradeoff function has the form

$$f(\vec{x}) = \sum_{p=1}^{n} \sum_{q=j}^{n} \sigma_{pq} x_p x_q + \sum_{p=1}^{n} \rho_p x_p \qquad 6.101$$

There is one parameter for each quadratic term, and one parameter for each linear term. The total number of parameters is $p = \dfrac{n(n+1)}{2} + n$. This is a general quadratic tradeoff function. The constraint 6.38 leads to the conditions

$$\rho_i > 0$$
$$\sigma_{ij} > 0$$

6.102

There are $\frac{n(n+3)}{2}$ of these conditions. We must have at least $k = \frac{n(n+3)}{2}$ pairs of points in order to find a solution for the parameters. Again, we minimize χ^2 with respect to each parameter.

Minimizing χ^2 with repsect to each parameter leads to the same problem from the previous section: the trivial solution $f(\vec{x}) = 0$. Again, we set the value of the tradeoff function at a specific point. This adds the constraint

$$f(\vec{x}) = \omega$$

6.103

Using the method of Lagrange, χ^2 is

$$\chi^2 = \sum_{l=1}^{k} [f(\vec{x}_{1l}) - f(\vec{x}_{2l})]^2 + 2\lambda[f(\vec{x}) - \omega]$$

6.104

Taking the partial derivatives,

$$\frac{\partial \chi^2}{\partial \sigma_{ij}} = \sum_{l=1}^{k} [f(\vec{x}_{1l}) - f(\vec{x}_{2l})][x_{1li}x_{1lj} - x_{2li}x_{2lj}] + \lambda \bar{x}_i \bar{x}_j = 0$$

6.105

$$\frac{\partial \chi^2}{\partial \rho_i} = \sum_{l=1}^{k} [f(\vec{x}_{1l}) - f(\vec{x}_{2l})][x_{1li} - x_{2li}] + \lambda \bar{x}_i = 0$$

6.106

$$\frac{\partial \chi^2}{\partial \lambda} = f(\vec{x}) - \omega = 0$$

6.107

or

$$\frac{\partial \chi^2}{\partial \sigma_{ij}} = \sum_{l=1}^{k} \left[\sum_{p=1}^{n} \sum_{q=j}^{n} \sigma_{pq} \left(x_{1lp}x_{1lq} - x_{2lp}x_{2lq} \right) \right.$$

$$\left. + \sum_{p=1}^{n} \rho_p \left(x_{1lp} - x_{2lp} \right) \right] \left[x_{1li}x_{1lj} - x_{2li}x_{2lj} \right]$$

$$+ \lambda \bar{x}_i \bar{x}_j = 0$$

6.108

$$\frac{\partial \chi^2}{\partial \rho_i} = \sum_{l=1}^{k} \left[\sum_{p=1}^{n} \sum_{q=j}^{n} \sigma_{pq} \left(x_{1lp} x_{1lq} - x_{2lp} x_{2lq} \right) \right.$$

$$\left. + \sum_{p=1}^{n} \rho_p \left(x_{1lp} - x_{2lp} \right) \right] \left[x_{1li} - x_{2li} \right] + \lambda \bar{x}_i = 0 \qquad \text{6.109}$$

$$\frac{\partial \chi^2}{\partial \lambda} = \left[\sum_{p=1}^{n} \sum_{q=j}^{n} \sigma_{pq} \left(x_{1lp} x_{1lq} - x_{2lp} x_{2lq} \right) \right.$$

$$\left. + \sum_{p=1}^{n} \rho_p \left(x_{1lp} - x_{2lp} \right) \right] - \omega = 0 \qquad \text{6.110}$$

These equations may be put in matrix form similar to the previous section. Separating the quadratic terms form the linear terms as we have done here actually increases the complication presenting the solution.

In the next section we solve a more general form. Rather than finishing the solution here, we use the results of the next section after we solve the more general form.

6.3-c GENERAL POLYNOMIAL TRADEOFF

Now we generalize the previous results. We assume the tradeoff function is a series of polynomial terms, each multiplied by an unknown coefficient. The dependent variables are on the range [0,1]. The tradeoff function has the form

$$f(\vec{x}) = \sum_{p=1}^{m} \rho_p g_p(\vec{x}) \qquad \text{6.111}$$

There is one parameter for each of the m terms in the tradeoff. Thus, we must have at least m pairs of points identified in order to find a solution.

The tradeoff function must also meet the Pareto constraint

$$\sum_{i \neq j}^{n} \frac{\partial f}{\partial x_i} \frac{\partial f}{\partial x_j} > 0 \qquad \text{6.112}$$

In addition, we set the value of the tradeoff to ω as some point $\vec{x}$:

$$f(\vec{x}) = \omega \qquad \text{6.113}$$

These conditions lead to the χ^2

$$\chi^2 = \sum_{l=1}^{k} [f(\vec{x}_{1l}) - f(\vec{x}_{2l})]^2 + 2\lambda [f(\vec{x}) - \omega] \tag{6.114}$$

Taking the partial derivatives

$$\frac{\partial \chi^2}{\partial \rho_i} = \sum_{l=1}^{k} [f(\vec{x}_{1l}) - f(\vec{x}_{2l})] \left[\frac{\partial f(\vec{x}_{1l})}{\partial \rho_i} - \frac{\partial f(\vec{x}_{2l})}{\partial \rho_i} \right] + \lambda \frac{\partial f(\vec{x})}{\partial \rho_i} = 0 \tag{6.115}$$

$$\frac{\partial \chi^2}{\partial \lambda} = f(\vec{x}) - \omega = 0 \tag{6.116}$$

Substituting the form of the tradeoff function,

$$\frac{\partial \chi^2}{\partial \rho_i} = \sum_{l=1}^{k} \sum_{p=1}^{m} \rho_p \left[g_p(\vec{x}_{1l}) - g_p(\vec{x}_{2l}) \right] [g_i(\vec{x}_{1l}) - g_i(\vec{x}_{2l})] + \lambda g_i(\vec{x}) = 0 \tag{6.117}$$

$$\frac{\partial \chi^2}{\partial \lambda} = \sum_{p=1}^{m} \rho_p g_p(\vec{x}) - \omega = 0 \tag{6.118}$$

In matrix form

$$\begin{bmatrix} \rho_1 \\ \rho_2 \\ \vdots \\ \rho_n \\ \lambda \end{bmatrix} = \begin{bmatrix} \mathcal{M}_{11} & \mathcal{M}_{12} & \cdots & \mathcal{M}_{1m} & g_1(\vec{x}) \\ \mathcal{M}_{21} & \ddots & & & g_2(\vec{x}) \\ \vdots & & \ddots & & \vdots \\ \mathcal{M}_{m1} & \mathcal{M}_{m2} & \cdots & \mathcal{M}_{mm} & g_m(\vec{x}) \\ g_1(\vec{x}) & g_2(\vec{x}) & \cdots & g_m(\vec{x}) & 0 \end{bmatrix}^{-1} \begin{bmatrix} 0 \\ 0 \\ \vdots \\ 0 \\ \omega \end{bmatrix} \tag{6.119}$$

where

$$\mathcal{M}_{ij} = \sum_{l=1}^{k} [g_i(\vec{x}_{1l}) - g_i(\vec{x}_{2l})][g_j(\vec{x}_{1l}) - g_j(\vec{x}_{2l})] \tag{6.120}$$

Generally, the solution obtained by taking the inverse of $\mathcal{M}$ is not required to meet the constraint from 6.112. Computing the matrix inverse is generally straightforward with numerical methods. Before embarking on a minimization scheme such as those from Chapter 2, it is usually worthwhile to check the solution from the matrix inverse to see if the constraint is met in the operational region of interest.

As an example of applying this, examine the generalized quadratic in two dimensions. Set the functions as

$$g_1(x, y) = x^2 \tag{6.121}$$

$$g_2(x, y) = xy \tag{6.122}$$

$$g_3(x, y) = y^2 \qquad \text{6.123}$$

$$g_4(x, y) = x \qquad \text{6.124}$$

$$g_5(x, y) = y \qquad \text{6.125}$$

For equivalent points, use the pairs of points in Table 6.

Point 1	Point 2
(.5, .7)	(.3, .8)
(.8, .6)	(.7, .7)
(.25, .4)	(.3, .3)
(.1, .6)	(.25, .55)
(.1, .6)	(.40, .47)
(.25, .55)	(.40, .47)
(.8, .2)	(.7, .3)
(.5, .4)	(.4, .55)
(.5, .1)	(.4, .15)

Table 6: Sample points identified as equivalent.

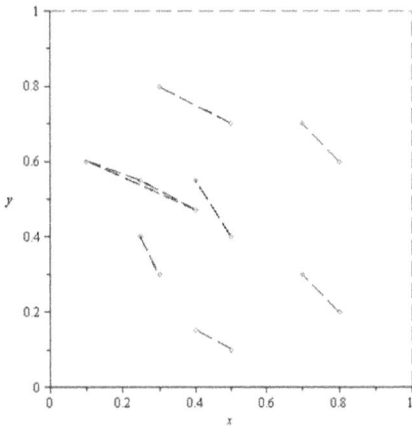

Figure 84: Sample points identified as equivalent.

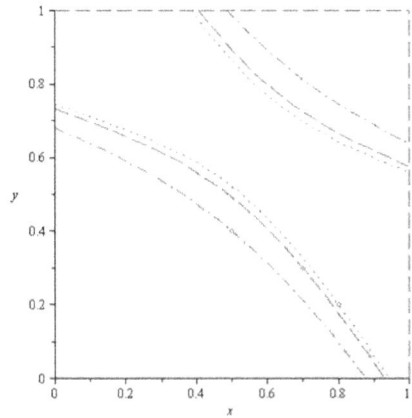

Figure 85: Points in the operation space showing sample fits.

These points are shown in Figure 84, and some sample fit curves are shown in Figure 85. The fit tradeoff function has negative slope at every point as expected. However, the fit is hyperbolic, and the two parts of the hyperbola are in the operational region. Although the value curves are negatively sloped for the curves shown, the value curve is not rational because the value curve equates points that are Pareto-comparable.

In general, the matrix equation from 6.119 may be used to obtain the best fit tradeoff function to the pairs of points identified by the user. The solution from the matrix equation must be compared to the requirement that the Pareto condition must be met everywhere. This constraint may not be met from strictly using the matrix equation. In these cases, the parameters need to be adjusted so that the tradeoff function is negative everywhere in the operational region.

If the matrix equations do not yield an appropriate solution, the optimization methods from Chapter 2 may be used to identify a best fit solution matching the slope constraints. This often requires numerical methods in order to identify suitable tradeoff functions.

6.4 Lagrange Multipliers with Errors

There may be errors associated with each operational point. The operational point $\vec{x}$ may have errors associated with each component. These are represented as $\delta\vec{x}$.

Uncertainty in the operational point leads to uncertainty in the value of the tradeoff function. Propagation of errors may be used to compute the error in the value of the tradeoff function when evaluated at the point $\vec{x}$. If the components of the operational point are independent, propagation of errors yields

$$\left(\delta f(\vec{x})\right)^2 = \sum_{i=1}^{n} \left(\frac{\partial f(\vec{x})}{\partial x_i}\right)^2 (\delta x_i)^2 \qquad 6.126$$

With this we can compute the uncertainty in the tradeoff function based on the error in the operational point.

This information may be incorporated into the χ^2 for the fit. If we know the uncertainty in the tradeoff function, the uncertainty in the difference between tradeoff functions is

$$(\delta[f(\vec{x}_{1l}) - f(\vec{x}_{2l})])^2 = \left(\delta f(\vec{x}_{1l})\right)^2 + \left(\delta f(\vec{x}_{2l})\right)^2 \qquad 6.127$$

Set

$$\delta(\vec{x}_{1l}, \vec{x}_{2l}) = \sqrt{\left(\delta f(\vec{x}_{1l})\right)^2 + \left(\delta f(\vec{x}_{2l})\right)^2} \qquad 6.128$$

The terms in χ^2 may be weighted by the uncertainty in each term:

$$\chi^2 = \sum_{l=1}^{k} \frac{[f(\vec{x}_{1l}) - f(\vec{x}_{2l})]^2}{\delta(\vec{x}_{1l}, \vec{x}_{2l})} + 2\lambda[f(\vec{\tilde{x}}) - \omega] \qquad \text{6.129}$$

There is no error associated with the second term because this term is meant to be exact. We want the value of the tradeoff function exactly at the point $\vec{\tilde{x}}$, so there is no uncertainty in the operational point in question.

Incorporating uncertainties into the fit is appropriate when the errors at different operational points are different. When the errors are all the same, this just multiplies each term in the first sum by a constant value. This does not affect the fit. However, when the errors are different (especially when the errors at some points are much less that the errors at other points), by incorporating the uncertainties into χ^2, the resulting fit better reflects our understanding of the operational scenario.

7 Continuous Dependent Functions

7.1 Continuous Operational Region

Thus far we have examined multiobjective optimization in a general sense. In this section we examine the characteristics of the multiobjective optimization problem particular to a continuous operational region.

If the operational space is a finite subset of $\mathbb{R}^N$, we can translate and scale the operational region to the unit cube $\mathbb{U}^N$. In this region, all dependent variables are on the range $[0,1]$. This transformation does not affect our analysis, but simplifies the resulting expressions and examples.

We also assume that the axes are oriented so that 'better' values are indicated by increasing values for the dependent variables. Again, this does not affect the analysis, but provides a consistent method for graphically displaying the results. Other orientations may be treated with similar methods.

A continuous operational region is a subset of $\mathbb{U}^N$. Each point in the operational region represents a set of dependent values that is operationally realizable. In other words, each point of the operational region is a particular set of dependent values that is achievable by the system under study.

Figure 86 shows an example of an operational region with the Pareto frontier highlighted, while Figure 87 shows the frontier by itself. The operational region represents the values for the dependent variables that are achievable. The Pareto frontier is the set of operational points that simultaneously maximize both dependent variables. In this example, the frontier is a continuous curve connecting the two extremes of the values for the dependent variables.

The frontier can be discontinuous as well. Figure 88 has an operational region where the Pareto frontier is discontinuous along the vertical axis. In this case, the operational region is concave, and this causes the discontinuity in the frontier. Figure 89 shows the frontier in isolation where is it much easier to see the discontinuity along the vertical axis.

Similarly, Figure 90 and Figure 91 provide an example of an operational region where the frontier is discontinuous along the horizontal axis. Again, the concave area of the operational region along the frontier leads to a discontinuity.

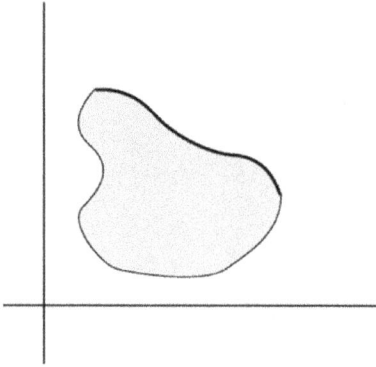

Figure 86: Example of an operational region with the Pareto frontier highlighted.

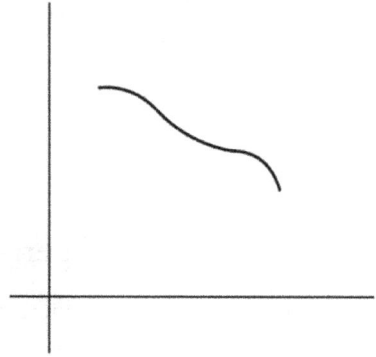

Figure 87: Previous figure showing only the Pareto frontier.

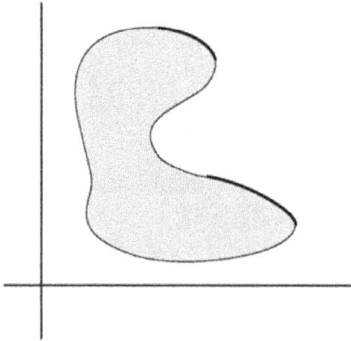

Figure 88: Example of an operational region with the frontier highlighted where the frontier has a discontinuity along the vertical axis.

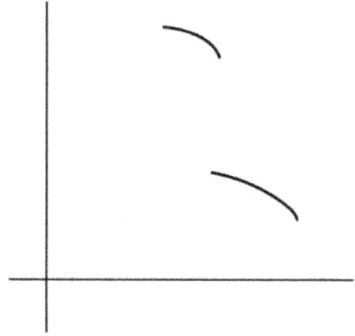

Figure 89: Previous figure showing only the Pareto frontier.

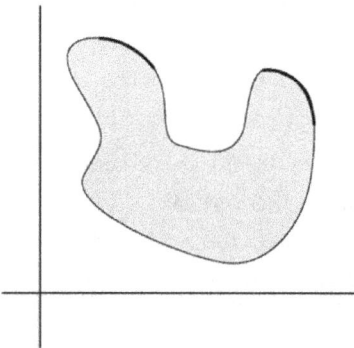

Figure 90: Example of an operational region with the frontier highlighted where the frontier has a discontinuity along the horizontal axis.

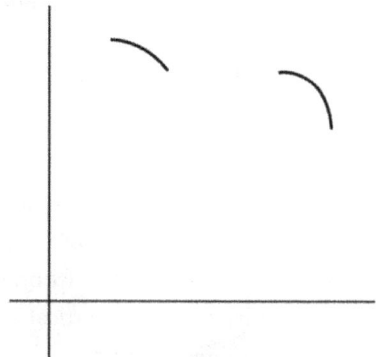

Figure 91: Previous figure showing only the Pareto frontier.

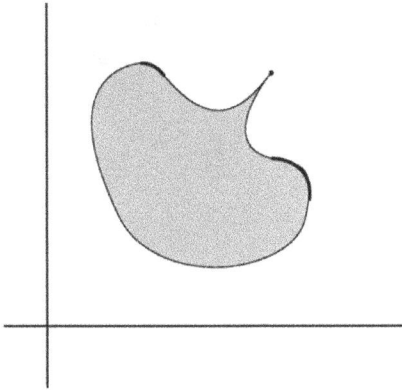

Figure 92: Example of an operational region with the Pareto frontier highlighted where the frontier has a discontinuity along both the x- and y-axes.

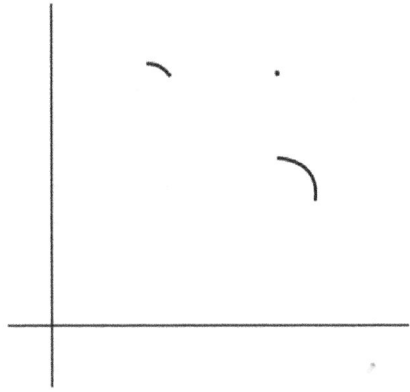

Figure 93: Previous figure showing only the Pareto frontier.

Finally, Figure 92 and Figure 93 provide an example of a discontinuity along both the vertical and horizontal axes simultaneously. When this happens, the corresponding section of the frontier is reduces to a single point.

There are many ways to specify the operational region. Some common methods of identifying the operational region are provided below.

Map of the Region – In proving mathematical theorems, it is common to assume the existence of a map $f: \mathbb{R}^N \to [0,1]$ that specifies if each point is or is not in the operational region. Such a map is useful as a mathematical concept, but may not be useful in practice.

Map of the Boundary – The region may be specified by identifying the boundary of the region. All points inside or on the boundary are considered part of the operational region. Mathematically, if the region is expressed as the map $f: \mathbb{R}^N \to [0,1]$, the boundary of the region is often designated ∂f.

Function of Dependent Variables for the Boundary – A common approach to specifying an operational region is to provide one or more functions of the dependent variables that identify the boundary of the operational region.

Parametric Expressions for the Region – In many practical cases, the region is expressed as a set of parametric equations for the dependent variables. Suppose we have N dependent variables, and each dependent variable is expressed as a function over M independent variables. Expressing the dependent variables in terms of the independent variables creates a set of parametric expressions for the points in the region.

For example, let an operational region in two dimensions have the dependent variables x and y. Let the dependent variables be written as functions of the independent variables σ and τ. The operational region is all points $(x(\sigma,\tau), y(\sigma,\tau))$ over the range of the dependent variables σ and τ. Specifically, let

$$x(\sigma,\tau) = \frac{\sigma + \tau}{2} \qquad\qquad 7.1$$

$$y(\sigma,\tau) = 1 - \frac{\sigma^2 + \tau^2}{2} \qquad\qquad 7.2$$

where $\sigma, \tau \in [0,1]$.

Equations 7.1 and 7.2 define a set of points that are part of the region of interest. The region is the set of points $(x(\sigma,\tau), y(\sigma,\tau))$ where $\sigma, \tau \in [0,1]$. This region is shown in Figure 94.

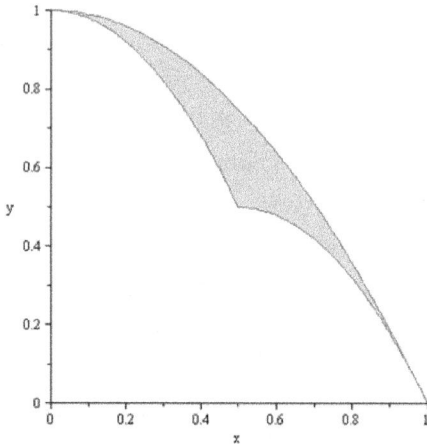

Figure 94: Region defined from the parametric equations 7.1 and 7.2.

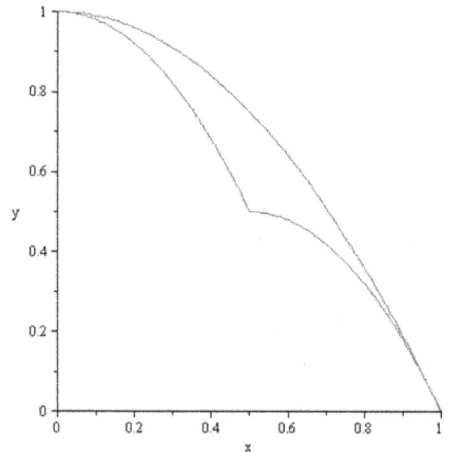

Figure 95: Boundary of the region defined from the parametric equations 7.1 and 7.2

Parametric Expressions for the Boundary – Another method for defining the region is to use parametric equations to define the boundary of the region. The boundary of the region is given by a set of parametric equations.

Continuing the example above, let an operational region in two dimensions have the dependent variables x and y, and let the dependent variables be written as functions of the independent variable τ. The parametric equations

$$I \qquad\qquad (x(\tau), y(\tau)) = \left(\frac{\tau}{2}, 1 - \frac{\tau^2}{2}\right) \qquad\qquad 7.3$$

| II | $$\big(x(\tau), y(\tau)\big) = \left(\frac{1+\tau}{2}, \frac{1-\tau^2}{2}\right)$$ | 7.4 |

| III | $$\big(x(\tau), y(\tau)\big) = (\tau, 1 - \tau^2)$$ | 7.5 |

where $\sigma, \tau \in [0,1]$.

Equations 7.3-7.5 define the boundary of the region of interest. This boundary is shown in Figure 95. The region is the set of points in the interior of the boundary.

Direct Expressions for the Boundary – The boundary may also be expressed as a set of equations for the dependent variables. From the previous example, examine the equations

| I | $$(x, y) = (x, 1 - 2x^2)$$ | 7.6 |

| II | $$(x, y) = \big(x, 2x(1 - x)\big)$$ | 7.7 |

| III | $$(x, y) = (x, 1 - x^2)$$ | 7.8 |

where $0 \le x \le 1$ in each equation.

These equations create the same boundary from the previous example. Thus, the boundary may be expressed as a set of equations involving the dependent variables.

Inequalities – A region may be expressed as a set of inequalities over the dependent variables. This is often another method to express the boundary. For example, the inequalities

| I | $$1 - 2x^2 \le y \le 1 - x^2$$ | 7.9 |

| II | $$2x(1 - x) \le y \le 1 - x^2$$ | 7.10 |

| III | $$0 \le x \le 1$$ | 7.11 |

define the same region as in the previous example. Essentially, the equations define the same boundary, and the inequalities express the understanding that the region is the set of points that are interior to the boundary.

Mixed Forms – The boundary or region may be expressed by combining the methods above. For example, consider the region defined by the equations

| I | $$x^2 + y^2 = \sigma^2$$ | 7.12 |

| II | $$\frac{1}{2} \le \sigma \le 1$$ | 7.13 |

III $0 \leq x \leq 1$ 7.14

IV $0 \leq y \leq 1$ 7.15

This region is shown in Figure 96. Essentially, this is the boundary

$$x^2 + y^2 = 1$$ 7.16

$$x^2 + y^2 = \frac{1}{4}$$ 7.17

where $0 \leq x, y \leq 1$. The above form is mixed because equation 7.12 related the dependent variables x and y with the parametric independent variable σ. This equation mixes the inequality method with the parametric method.

As another example, examine the region defined from

I $y = 1 - \sigma x^3$ 7.18

II $1 \leq \sigma \leq 10$ 7.19

III $0 \leq x \leq 1$ 7.20

IV $0 \leq y \leq 1$ 7.21

This region is shown in Figure 97. Again, the first equation mixes the form of a dependent variable equation with a parametric equation. Similar to the previous example, this defines a region as the interior of the boundary formed from the extreme values of the independent variable.

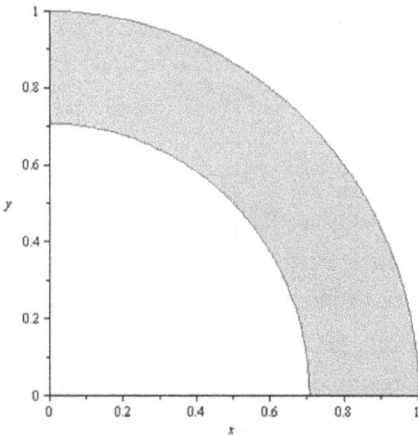

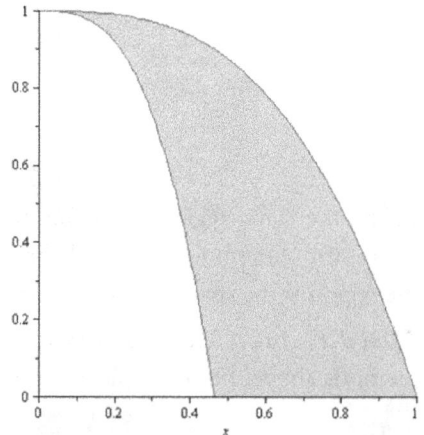

Figure 96: Region defined from the mixed equations 7.12-7.15.

Figure 97: Boundary of the region defined from the parametric equations 7.1 and 7.2

7.2 Boundary Turning Points

In this section we examine turning points on the boundary. Turning points are used in the next section to identify the entire frontier in two dimensions. In this section we identify methods for computing the turning points in two dimensions.

Figure 98-Figure 101 provides examples of continuous operational regions and their associated frontiers. In each case, the endpoint of a section of the frontier coincides with a local minimum or maximum with respect to one of the dependent variables.

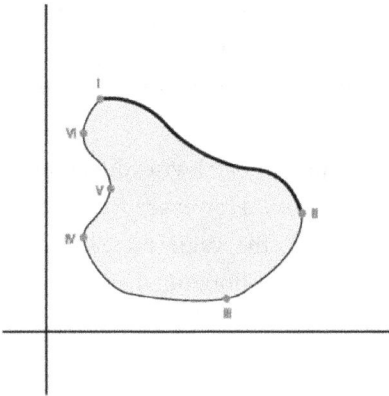

Figure 98: Example of an operational region with the Pareto frontier highlighted and boundary turning points indicated.

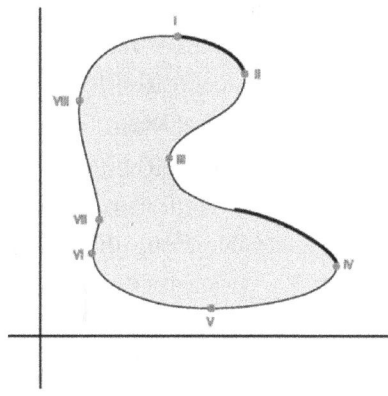

Figure 99: Example of an operational region with the frontier highlighted where the frontier has a discontinuity along the vertical axis.

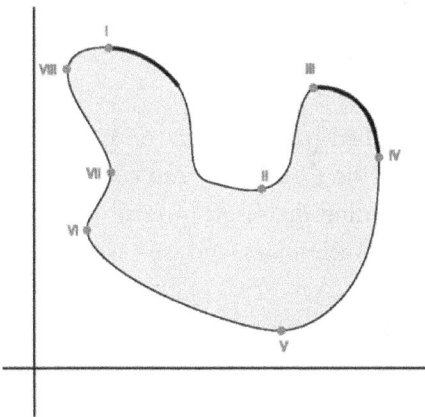

Figure 100: Example of an operational region with the frontier highlighted where the frontier has a discontinuity along the horizontal axis.

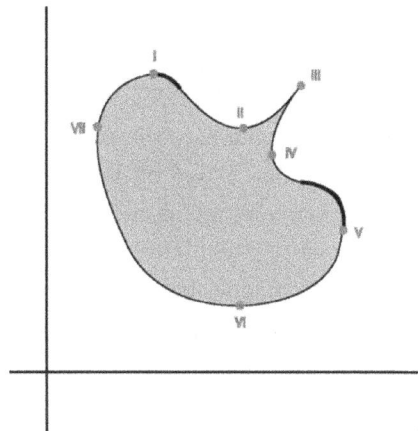

Figure 101: Example of an operational region with the Pareto frontier highlighted where the frontier has a discontinuity along both axes.

For example, in Figure 98, the frontier is a single continuous section. The left endpoint of the frontier coincides with the maximum value for the vertical axis. The right endpoint of the frontier coincides with the maximum value for the vertical axis.

The points that are local maxima or minima with respect to one of the dependent variables are called turning points. Identifying the turning points on the frontier is helpful in identifying the points on the frontier.

For example, there are six turning points on the boundary of the region in Figure 98. The first turning point (labeled I) is the maximum value along the vertical axis. Imaging projecting the boundary onto the vertical axis. As we traverse the boundary clockwise approaching I, the value on the vertical axis increases. As we pass through I, the value on the vertical axis reaches a maximum, then decreases. In this respect, point I represents a maximum with respect to the vertical axis.

As we continue to travel along the boundary in the clockwise direction, the value along the vertical axis continues to decrease. However, we may also examine the value along the horizontal axis. Here, the value increases as we approach II. As we pass through II, the value along the horizontal axis reaches a maximum, then begins to decrease. At the same time, the value along the vertical axis continually decreases as we pass through II. Thus, II is a maximum with respect to the horizontal axis.

Continuing along the boundary, we reach a minimum for the vertical axis at point III, a minimum for the horizontal axis at IV, a local maximum for the horizontal axis at V, and a local minimum for the horizontal axis at VI.

These six points are the turning points for the boundary. Each turning point is an extrema with respect to at least one of the dependent variables. Additional examples of turning points are provided in Figure 99-Figure 101.

Suppose the boundary of the region is specified parametrically or as a set of direct equations. Based on the expression for the frontier, we can compute the turning points by taking the derivative and setting the result to zero. Different expressions for the boundary require different techniques, and we examine two common techniques below.

Parametric Expressions – Suppose the boundary for the region is expressed parametrically as $(x(\tau), y(\tau))$ where $0 \le \tau < 1$. The turning points are the values of τ where either

$$\frac{dx(\tau)}{d\tau} = 0 \qquad\qquad 7.22$$

or

$$\frac{dy(\tau)}{d\tau} = 0 \qquad\qquad 7.23$$

We identify the turning points by identifying every value τ_k that satisfies one of the two equations above. The turning points are the points $(x(\tau_k), y(\tau_k))$.

As an example, examine the simple parametric forms

$$x(\tau) = \frac{1}{2}(1 + \cos 2\pi\tau) \qquad\qquad 7.24$$

$$y(\tau) = \frac{1}{2}(1 + \sin 2\pi\tau) \qquad\qquad 7.25$$

where $0 \le \tau < 1$.

The region is the circle shown in Figure 102. The turning points are found by computing the derivatives of the dependent variables with respect to the independent parameter:

$$\frac{dx(\tau)}{d\tau} = \pi \sin 2\pi\tau = 0 \qquad\qquad 7.26$$

$$\frac{dy(\tau)}{d\tau} = \pi \cos 2\pi\tau = 0 \qquad\qquad 7.27$$

This leads to the four values $\tau = 0, \frac{1}{4}, \frac{1}{2}, \frac{3}{4}$. The turning points are:

$$(x, y) = \begin{cases} \left(1, \frac{1}{2}\right) & \left(0, \frac{1}{2}\right) \\ \left(\frac{1}{2}, 1\right) & \left(\frac{1}{2}, 0\right) \end{cases} \qquad\qquad 7.28$$

As another example, consider the cardioid region

$$x(\tau) = \frac{2}{5} + \frac{1}{8}(2\cos(2\pi\tau) - \cos(4\pi\tau)) \qquad\qquad 7.29$$

$$y(\tau) = \frac{2}{5} + \frac{1}{8}(2\sin(2\pi\tau) - \sin(4\pi\tau)) \qquad\qquad 7.30$$

The turning points are given by the values of τ where either $x(\tau)$ or $y(\tau)$ is zero. Taking the derivatives,

$$\frac{dx(\tau)}{d\tau} = -\frac{\pi}{2}(\sin(2\pi\tau) - \sin(4\pi\tau)) = 0 \qquad\qquad 7.31$$

The zeros are at

$$\frac{dx(\tau)}{d\tau} = 0 \Rightarrow \tau = 0, \frac{1}{6}, \frac{1}{2}, \frac{5}{6} \qquad 7.32$$

Similarly for y:

$$\frac{dy(\tau)}{d\tau} = \frac{\pi}{2}(\cos(2\pi\tau) - \cos(4\pi\tau)) = 0 \qquad 7.33$$

$$\frac{dy(\tau)}{d\tau} = 0 \Rightarrow \tau = 0, \frac{1}{3}, \frac{2}{3} \qquad 7.34$$

Combining these results, the turning points are located at

$$\tau = 0, \frac{1}{6}, \frac{1}{3}, \frac{1}{2}, \frac{2}{3}, \frac{5}{6} \qquad 7.35$$

The cardioid region and the turning points are shown in Figure 103.

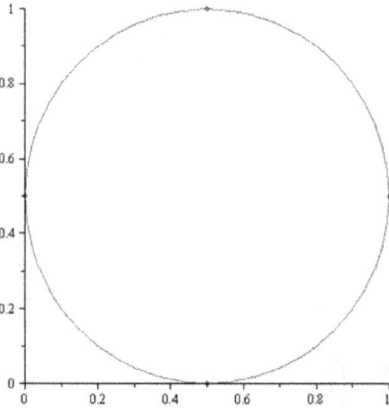

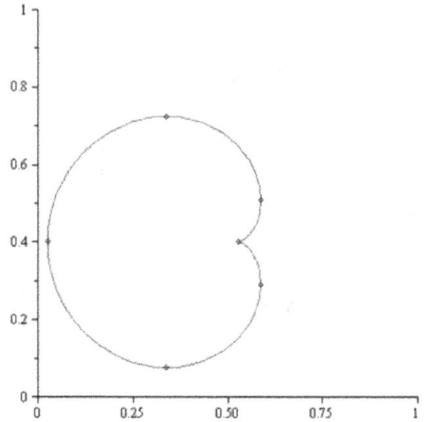

Figure 102: Circular operational region with turning points indicated.

Figure 103: Cardioid operational region with turning points indicated.

Direct Equations – The boundary may be identified as set of equations directly relating the dependent variables. For example, an elliptical region may be expressed as the single equation

$$4(x - .25)^2 + 10(y - .5)^2 = .25 \qquad 7.36$$

We need to find the extrema with respect to each axis. First, take the differential of the equation:

$$8(x - .25)dx + 10(y - .5)dy = 0 \qquad 7.37$$

Dividing by dx,

$$8(x - .25) + 10(y - .5)\frac{dy}{dx} = 0 \qquad \text{7.38}$$

$$\frac{dy}{dx} = -\frac{4(x - .25)}{5(y - .5)} \qquad \text{7.39}$$

This is the derivative along the vertical axis. The derivative is zero when $x = .25$. The turning points are

$$\frac{dy}{dx} = 0 \Rightarrow (.25, .3419), (.25, .6581) \qquad \text{7.40}$$

The corresponding values for y in the above expression are determined by substituting $x = .25$ into 7.36 and solving for y.

Returning to 7.37, if we divide by dy,

$$8(x - .25)\frac{dx}{dy} + 10(y - .5) = 0 \qquad \text{7.41}$$

$$\frac{dx}{dy} = -\frac{5(y - .5)}{4(x - .25)} \qquad \text{7.42}$$

This is the derivative along the horizontal axis. The extremum are at the point $y = .5$ with the corresponding turning points

$$\frac{dx}{dy} = 0 \Rightarrow (0, .5), (.5, .5) \qquad \text{7.43}$$

Thus, there are four turning points:

$$\begin{array}{cc} (.25, .3419) & (.25, .6581) \\ (0, .5) & (.5, .5) \end{array} \qquad \text{7.44}$$

as indicated in Figure 104.

As a final example, consider the region bounded by the parabolas

$$y = 1 - x^2 \qquad \text{7.45}$$

$$y = (x - 1)^2 \qquad \text{7.46}$$

The turning points along the vertical axis are:

$$\frac{dy}{dx} = 0 \Rightarrow \begin{array}{l} -2x = 0 \\ 2(x - 1) = 0 \end{array} \qquad \text{7.47}$$

These correspond to the points

$$(1, 0) \quad (0, 1) \qquad \text{7.48}$$

This region and the turning points are shown in Figure 105.

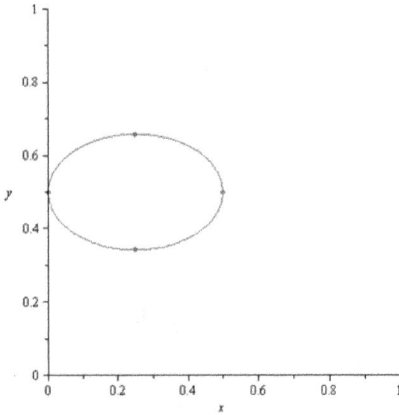

Figure 104: Elliptical operational region with turning points indicated.

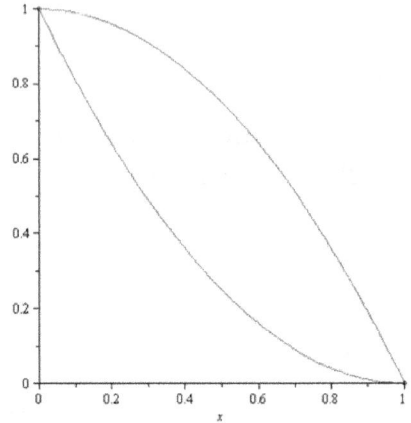

Figure 105: Parabolic operational region with turning points indicated.

7.3 Continuous Pareto Frontier

Turning points may be used to identify the Pareto frontier in two dimensions. Examining Figure 108-Figure 111, in many cases a segment of the frontier begins or ends on a turning point. We quantify this observation below.

If the operational region is two-dimensional and finite with a continuous, non-self-intersecting frontier, then there must be a maximum value for the vertical axis. There may be multiple points that correspond to the maximum value for the vertical axis. If so, each point has the same value for the vertical axis and different values for the horizontal axis. Thus, there is a unique point that corresponds to the maximum value for the vertical axis that Pareto dominates (this is the point with the largest value for the horizontal axis).

Similarly, there is a unique point that corresponds to the maximum value for the horizontal axis. Again, if there are multiple points with the maximum value for the horizontal axis, then one dominates the rest. The unique dominant point for the vertical axis may be different than the unique dominant point for the horizontal axis.

If we traverse the boundary in the clockwise direction, the Pareto frontier must begin on the dominant vertical point, and the frontier must end on the dominant horizontal point. Thus must be the case because the frontier must be negatively sloped. Hence, the frontier must begin at a high value for the vertical axis and move toward increasing values on the horizontal axis.

Moreover, the dominant maximum point on the vertical axis must be the first point on the frontier. This point dominates all points to the left because all of these points have a smaller value along the horizontal axis (points to the left), and cannot have a greater vertical value (this is the maximum vertical value). Hence, no prior point to the left can be on the frontier.

Finally, this point must be on the frontier because there is no point with a larger vertical value. Since no other point has a larger vertical value, there is no point that dominates this point, so this point must be on the frontier. Since there are no other points on the frontier to the left of this point, and since this point must be on the frontier, this must be the first point on the frontier.

Similar reasoning as the above may be employed to show that the dominant horizontal point must be the last point on the frontier. Thus, these two turning points provide the beginning and ending of the Pareto frontier.

If there are no turning points along the frontier between these, then all points between these are on the frontier. Suppose we have two consecutive turning points (consecutive as we traverse the boundary) that are both on the frontier. Let $\mathcal{B}$ be the set of points on the boundary between these turning points. Since there are no turning points between these, then the slope of the boundary is negative at every point in $\mathcal{B}$. But this means that all points in $\mathcal{B}$ are incomparable with each other.

Since no point in $\mathcal{B}$ dominates any other point in $\mathcal{B}$, all of these points must be on the frontier unless there exists another point not in $\mathcal{B}$ that dominates. Suppose there exists a point $p \notin \mathcal{B}$ on the frontier that dominates some point in $\mathcal{B}$. The point p must be in both the horizontal and vertical region between the turning points (otherwise p would dominate one of the endpoints, contradicting the assumption that both of these points are on the frontier).

Since the boundary is continuous, and since $p \notin \mathcal{B}$, there must be some path from one of the endpoints of $\mathcal{B}$ along the boundary that connects to p. Let $\mathfrak{R}$ be the region bounded by $\mathcal{B}$ and the point m constructed from the maximum horizontal and vertical values of $\mathcal{B}$. These points and regions are shown in Figure 106.

Since p dominates a point in $\mathcal{B}$ but does not dominate either of the endpoints of $\mathcal{B}$, the p must be in $\mathfrak{R}$. If p is in $\mathfrak{R}$, then the boundary must cross into $\mathfrak{R}$ at some point other than the endpoints. However, if the boundary crosses into $\mathfrak{R}$, then the point where the boundary intersects $\mathfrak{R}$ dominates one of the endpoints, contradicting the initial assumption that the endpoints are on the frontier.

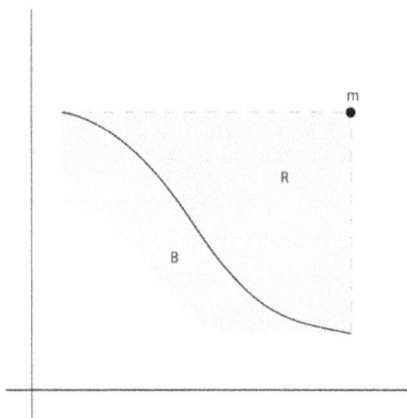

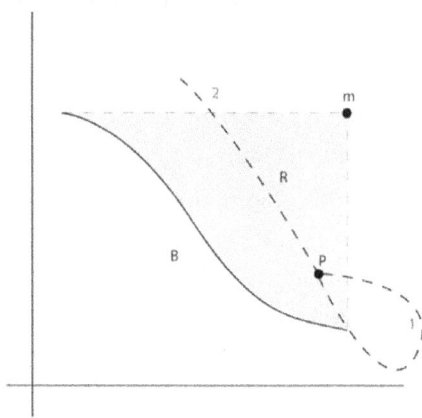

Figure 106: $\mathcal{B}$ is the set of points on the boundary of the region between consecutive turning points where the turning points are on the frontier. The region $\mathfrak{R}$ is the set of points between $\mathcal{B}$ and the point m. The point m is constructed from the maximum horizontal and vertical values of $\mathcal{B}$.

Figure 107: If the point p dominates some point in $\mathcal{B}$ but does not dominate the endpoints of $\mathcal{B}$, then p must lie in $\mathfrak{R}$. However, the boundary must connect p to the endpoints. If the boundary leaves $\mathfrak{R}$, then in order to get to p the boundary must cross either the horizontal or vertical lines. In this case one of the endpoints must be dominated by another point and hence cannot be on the frontier.

All of this is a complicated way to say that if the boundary is continuous and doesn't intersect itself, and if we identify two turning consecutive turning points that are both on the frontier, then all points on the boundary between these turning points are also on the frontier.

If only one turning point is on the frontier, then we may have a situation such as point III Figure 109 that is an isolated point on the frontier. However, in many cases, this turning point is the beginning of a section on the frontier that does not end at another turning point. Examples of this situation is shown in Figure 107 and Figure 108.

For example, point IV in Figure 107 is the dominant maximum along the horizontal axis. A portion of the boundary between points III and IV are actually on the frontier. However, midway along the boundary the points change from being on the frontier to not on the frontier. This is due to the presence of point II that dominates part of this section.

This is incorporated into the identification of the frontier by recognizing that turning point II dominates point III. We compute the intersection of the boundary with the vertical line drawn through point II. Where this line intersects the boundary between points III and IV provides the location of the cutoff between boundary points that are and are not on the frontier.

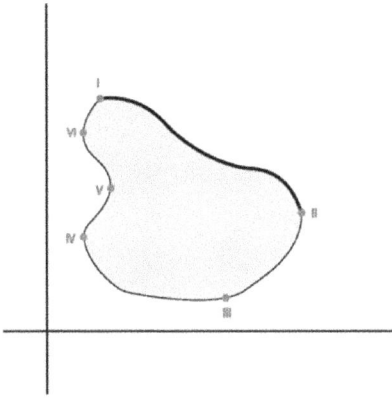

Figure 108: Example of an operational region with the Pareto frontier highlighted and boundary turning points indicated.

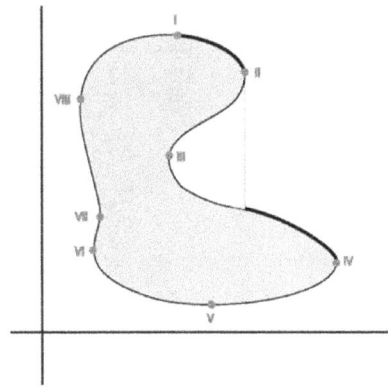

Figure 109: Example of an operational region with the frontier highlighted where the frontier has a discontinuity along the vertical axis. The vertical line indicated the portion of the boundary dominated by turning point II.

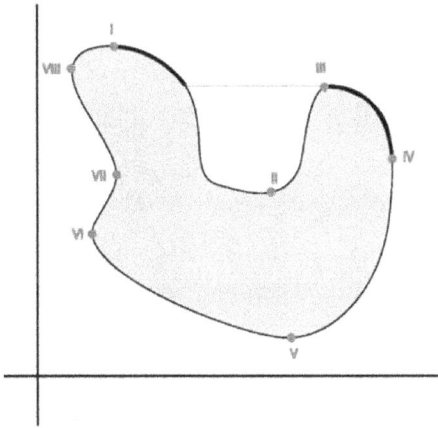

Figure 110: Example of an operational region with the frontier highlighted where the frontier has a discontinuity along the horizontal axis. The horizontal line indicated the portion of the boundary dominated by turning point III.

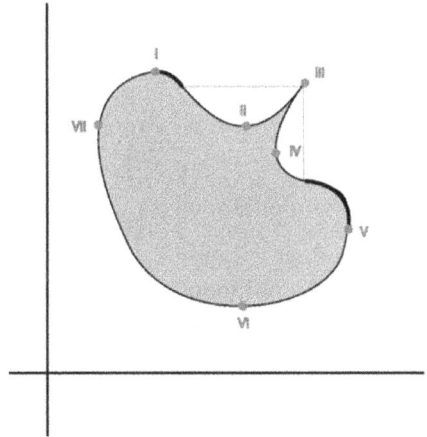

Figure 111: Example of an operational region with the Pareto frontier highlighted where the frontier has a discontinuity along both axes. The vertical and horizontal lines indicate the portion of the boundary dominated by turning point III.

As an example of allying this, consider again the cardioid region from Figure 103. Previously we found the turning points were located at

Combining these results, the turning points are located at

$$\tau = 0, \frac{1}{6}, \frac{1}{3}, \frac{1}{2}, \frac{2}{3}, \frac{5}{6} \qquad \text{7.49}$$

In this parameterization, increasing values for τ traverse the boundary in the counter-clockwise direction. The points on the boundary corresponding to these are

$$\tau_1 = 0 \quad (.5250, .4000) \tag{7.50}$$

$$\tau_2 = \frac{1}{6} \quad (.5875, .5083) \tag{7.51}$$

$$\tau_3 = \frac{1}{3} \quad (.3375, .7248) \tag{7.52}$$

$$\tau_4 = \frac{1}{2} \quad (.0250, .4000) \tag{7.53}$$

$$\tau_5 = \frac{2}{3} \quad (.3375, .0752) \tag{7.54}$$

$$\tau_6 = \frac{5}{6} \quad (.5875, .2917) \tag{7.55}$$

The dominant vertical point is the point τ_3 (highest y-value), and the dominant horizontal point is τ_2 (highest x-value, tied with τ_6, but the y-component of τ_2 is larger).

Thus, τ_2 begins the frontier while τ_3 ends the frontier. Since there are no turning points in between, all points on the boundary between τ_2 and τ_3 must also be on the frontier.

This example shows how the frontier may be determined from the turning points. By examining the turning points, we can determine which boundary points are on the frontier.

Although this technique is straightforward, it relies on two assumptions that may not be present in every situation. First, we must be able to identify the turning points. If the frontier is specified as a set of direct equations or parametrically, numeric techniques can often be employed to determine the turning points.

However, in some cases the boundary may not be specified as a simple set of equations that can be differentiated. In these situations, identifying the turning points may be complicates or often intractable.

Second, we assume that the boundary itself is identified. Again, when the boundary is specified by a set of equations, this assumption is fairly safe. However, when the boundary is not specified in this manner, even finding the points on the boundary may be difficult. We examine techniques for identifying the frontier in such situations in the next chapter.

8 Multiobjective Optimization Techniques

8.1 Preliminaries

The multiobjective optimization problem may generally be formulated as the maximization or minimization of a vector-valued function $\vec{F}$ subject to some equality and inequality constraints:

$$max \, \vec{F}(\vec{x}) = \begin{bmatrix} f_1(\vec{x}) \\ f_2(\vec{x}) \\ \vdots \\ f_n(\vec{x}) \end{bmatrix}$$

$$g_1(\vec{x}) = 0$$
$$g_2(\vec{x}) = 0$$
$$\vdots$$
$$g_k(\vec{x}) = 0$$
$$h_1(\vec{x}) \leq 0$$
$$h_2(\vec{x}) \leq 0$$
$$\vdots$$
$$h_l(\vec{x}) \leq 0$$

8.1

Multiobjective optimization attempts to maximize the n-dimensional vector function $\vec{F}$ subject to k equality constraints and l inequality constraints. Each of the components and the constraints are functions of the m-dimensional vector $\vec{x}$.

Multiobjective optimization problems are characterized by the dimensionality of the function under optimization as well as the dimensionality of the underlying vector space. A multiobjective optimization problem where $\vec{F}$ is a vector valued function on n-dimensions, and where the underlying parameter space for $\vec{x}$ has dimension m is called a $n \times m$-dimensional multiobjective optimization problem.

The space formed from allowable $\vec{F}$ is called the objective space, while the space formed from allowable $\vec{x}$ is called the parameter space. Multiobjective problems subject to constraints such as $g(\vec{x})$ and/or $h(\vec{x})$ above are called constrained multiobjective problems. Alternatively, multiobjective problems that are not subject to constraints are called unconstrained problems.

8.2 Turning Point Method

The turning point method computes the minima and maxima of each of the components of the objective function $\vec{F}$ and computed the Pareto frontier based on these anchor points. This method works on unconstrained $2x1$-dimensional multiobjective optimization problems and is discussed in detail in section 7.3.

Let the objective function be

$$\vec{F}(x) = \begin{bmatrix} f_1(x) \\ f_2(x) \end{bmatrix}$$
8.2

Here, $\vec{F}$ is a two-component objective function over a single continuous variable x. The turning points are found by optimizing each component of $\vec{F}$. Let $x_i^1 \in X^1$ be the ith local optimum for $f_1(x)$ and $x_i^2 \in X^2$ be the ith local optimum for $f_2(x)$. Let $\bar{x}_i \in \bar{X} = X_1 \cup X_2$. The vectors $\vec{F}(\bar{x}_i)$ are the turning points for $\vec{F}$.

Order the turning points $\bar{x}_i$ in consecutive order so $\bar{x}_1 < \bar{x}_2 < \bar{x}_3 ... < \bar{x}_p$. Examine the value of the objective function $\vec{F}(x)$ at each turning point. Identify each point $\bar{x}_i$ where the corresponding $\vec{F}(\bar{x}_i)$ is not dominated by any other $\vec{F}(\bar{x}_j)$. Each of these $\vec{F}(\bar{x}_i)$ are points on the Pareto frontier.

Examine the sequence of consecutive points $\bar{x}_i$ where $\vec{F}(\bar{x}_i)$ is on the Pareto frontier. If the next consecutive point $\bar{x}_{i+1}$ has $\vec{F}(\bar{x}_{i+1})$ is also on the Pareto frontier, then all points $\vec{F}(x)$ where $x \in [\bar{x}_i, \bar{x}_{i+1}]$ are also on the Pareto frontier. If $\vec{F}(\bar{x}_i)$ dominates $\vec{F}(\bar{x}_{i+1})$, then none of the points $\vec{F}(x)$ where $x \in [\bar{x}_i, \bar{x}_{i+1}]$ are on the frontier.

If $\vec{F}(\bar{x}_{i+1})$ is not on the frontier, but $\vec{F}(\bar{x}_i)$ does not dominate $\vec{F}(\bar{x}_{i+1})$, then skip ahead to the next consecutive point $\bar{x}_j$ where $\vec{F}(\bar{x}_j)$ is on the Pareto frontier. Since both $\vec{F}(\bar{x}_i)$ and $\vec{F}(\bar{x}_j)$ are on the frontier, then neither point dominated the other. Thus, either

I
$$f_1(x_i) > f_1(x_j)$$
$$f_2(x_i) < f_2(x_j)$$
8.3

or

II
$$f_1(x_i) < f_1(x_j)$$
$$f_2(x_i) > f_2(x_j)$$
8.4

Under Case I, locate the minimum $x_b > x_i$ where $f_1(x_b) = f_1(x_j)$. All points $\vec{F}(x)$ where $x \in [\bar{x}_i, x_b]$ are on the Pareto frontier. Under Case II, locate the minimum $x_b > x_i$ where $f_2(x_b) = f_2(x_j)$. All points $\vec{F}(x)$ where $x \in [\bar{x}_i, x_b]$ are on the Pareto frontier.

The above process may be carried out for the point previous to $\bar{x}_i$. In this way we identify the points on the frontier connected to $\bar{x}_i$. The entire process may be repeated for each non-dominated point $\bar{x}_i$ to obtain the entire frontier.

As an example of the turning point method, consider the unconstrained multiobjective optimization problem

$$\vec{F}(t) = \left[\frac{(1-t)(.5 + .25\, sin(10\pi t))}{t} \right] \qquad 8.5$$

First compute the optima for $(1-t)(.5 + .25\, sin(10\pi t))$. The optima are found from the solution to the equation

$$\frac{\partial f_1(t)}{\partial t} = -.5 - .25\, sin(10\pi t) + 2.5\pi(1-t)\, cos(10\pi t) = 0 \qquad 8.6$$

Solutions to this equation may be found using the root finding techniques discussed in chapter 2. There are ten such optima, five minima and five maxima as shows in the graph of the objective function in Figure 112. Further, there are no optima from $f_2(t)$.

Examining the optima in sequence, each non-dominated optima falls into Case II in step S-8 from the algorithm below. In each case, we can again use the root finding techniques from chapter 2 to identify the points x_b in S-8.

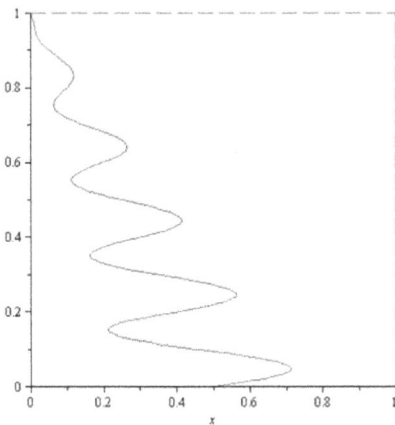

Figure 112: Multiobjective optimization example for a 2 × 1 dimensional objective.

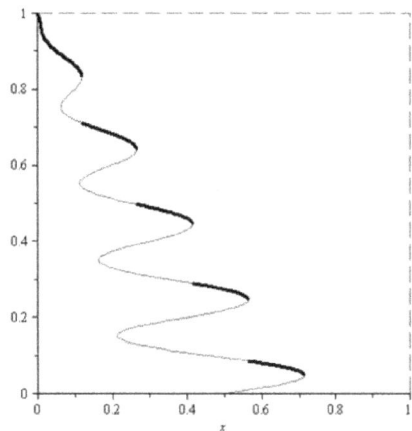

Figure 113: Pareto frontier highlighted for the example problem.

Turning Point Method

I-1   Independent variable x

        Objective function

I-2

$$\vec{F}(x) = \begin{bmatrix} f_1(x) \\ f_2(x) \end{bmatrix}$$

S-1   Compute all points x where $\dfrac{\partial f_1(x)}{\partial x} = 0$ and $\dfrac{\partial^2 f_1(x)}{\partial x^2} \neq 0$ (the point x_i is either a minimum or maximum, not a point of inflection). Designate the set of all such points as the set X_1.

S-2   Compute all points x where $\dfrac{\partial f_2(x)}{\partial x} = 0$ and $\dfrac{\partial^2 f_2(x)}{\partial x^2} \neq 0$. Designate the set of all such points as the set X_2.

S-3   Compute the union of the sets $\bar{X} = X_1 \cup X_2$

S-4   Order the points $\bar{x}_i \in \bar{X}$ so that $\bar{x}_1 < \bar{x}_2 < \cdots < \bar{x}_p$

S-5   Compute $\vec{F}(\bar{x}_i)$ for each $\bar{x}_i \in \bar{X}$.

S-6   Identify each $\bar{x}_i$ where the corresponding $\vec{F}(\bar{x}_i)$ is not dominated by any other $\vec{F}(\bar{x}_j)$

S-7   Find the first $\bar{x}_i$ with a non-dominated $\vec{F}(\bar{x}_i)$

        Examine $\bar{x}_{i+1}$:

           i.   If $\vec{F}(\bar{x}_i)$ dominates $\vec{F}(\bar{x}_{i+1})$, then none of the points $\vec{F}(x)$ where $\bar{x}_i \leq x \leq \bar{x}_{i+1}$ are on the Pareto frontier. Continue from S-9.

           ii.   If $\vec{F}(\bar{x}_{i+1})$ is non-dominated, then all of the points $\vec{F}(x)$ where $\bar{x}_i \leq x \leq \bar{x}_{i+1}$ are on the Pareto frontier. Continue from S-9.

S-8   iii.   Otherwise, find the next sequential point $\bar{x}_j$ where $\vec{F}(\bar{x}_j)$ is non-dominated.

             a.   Case I:

$$f_1(x_i) > f_1(x_j)$$
$$f_2(x_i) < f_2(x_j)$$

               Find the minimum $x_b > x_i$ where $f_1(x_b) = f_1(x_j)$. All points $\vec{F}(x)$ where $\bar{x}_i \leq x \leq \bar{x}_b$ are on the Pareto frontier. Continue from S-9.

    b. Case II:

$$f_1(x_i) > f_1(x_j)$$
$$f_2(x_i) < f_2(x_j)$$

Find the minimum $x_b > x_i$ where $f_1(x_b) = f_1(x_j)$. All points $\vec{F}(x)$ where $\bar{x}_i \leq x \leq \bar{x}_b$ are on the Pareto frontier. Continue from S-9.

Examine $\bar{x}_{i-1}$:

    iv. If $\vec{F}(\bar{x}_i)$ dominates $\vec{F}(\bar{x}_{i-1})$, then none of the points $\vec{F}(x)$ where $\bar{x}_{i-1} \leq x \leq \bar{x}_i$ are on the Pareto frontier. Continue from S-9.

    v. If $\vec{F}(\bar{x}_{i+1})$ is non-dominated, then all of the points $\vec{F}(x)$ where $\bar{x}_{i-1} \leq x \leq \bar{x}_i$ are on the Pareto frontier. Continue from S-9.

    vi. Otherwise, find the previous sequential point $\bar{x}_j$ where $\vec{F}(\bar{x}_j)$ is non-dominated.

    c. Case I:

S-9

$$f_1(x_i) > f_1(x_j)$$
$$f_2(x_i) < f_2(x_j)$$

Find the minimum $x_b < x_i$ where $f_1(x_b) = f_1(x_j)$. All points $\vec{F}(x)$ where $\bar{x}_b \leq x \leq \bar{x}_i$ are on the Pareto frontier. Continue from S-9.

    d. Case II:

$$f_1(x_i) > f_1(x_j)$$
$$f_2(x_i) < f_2(x_j)$$

Find the minimum $x_b < x_i$ where $f_1(x_b) = f_1(x_j)$. All points $\vec{F}(x)$ where $\bar{x}_b \leq x \leq \bar{x}_i$ are on the Pareto frontier. Continue from S-9.

S-10   Repeat from S-7 using the next non-dominated point

O-1   The Pareto frontier is identified in the steps of the algorithm.

Algorithm XXXII: The turning point method may be used to identify the Pareto frontier.

8.3 Monte Carlo

The Monte Carlo method chooses points in the parameter space at random and computes the value of the vector in the operational space. Each point is compared to the current list of points on the Pareto frontier to determine if it is dominated by another point. If it is not dominated, the point is added to the list of points on the frontier.

The Monte Carlo algorithm works on problems of all dimensions. Any $n \times m$-dimensional multiobjective optimization problem may be approached with the Monte Carlo method. Furthermore, as the number of iterations of the algorithm increases, so does the number of points identified as candidates for the frontier.

There are two significant drawbacks for the Monte Carlo algorithm. First, each candidate point needs to be tested against each point currently identified as non-dominated. If a point on the non-dominated list dominates the test point, the testing is terminated and the candidate point discarded. However, in order to put the candidate point on the non-dominated list, the candidate point must be checked against each point in the list. As the number of points on the list increases, so does the run time of the algorithm.

Another drawback of the Monte Carlo method is the random selection of the points in parameter space. The mapping of points in parameter space to the objective space is rarely uniform. Thus, selecting points uniformly in the parameter space often leads to a biased representation in the objective space. Because of this, it may be difficult to identify points in the objective space that are near the true Pareto frontier.

As an example, consider the multiobjective optimization problem

$$\vec{F}(t) = \begin{bmatrix} t \\ (1-t)s \end{bmatrix}$$
$$\text{8.7}$$

where $t, s \in [0,1]$. A graph of the objective space is provided in Figure 114. The objective space is a triangular region, and the desire is to maximize the objective function. The Pareto frontier is the hypotenuse of the triangle.

The Monte Carlo method uses randomly generated points from the parameter space. The map of these points onto the objective space results in a non-uniform representation in the objective space. Figure 114 is colored to reflect this. Lighter areas have higher probability than darker areas.

Figure 115 shows the Pareto frontier obtained after 1000 iterations of the Monte Carlo algorithm. The algorithm identifies many points in the neighborhood of the frontier. As expected, there are more points identified near the frontier in the denser regions.

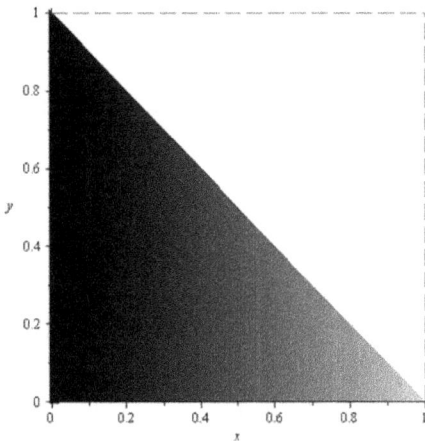

Figure 114: Multiobjective optimization example for a 2 × 2 dimensional objective.

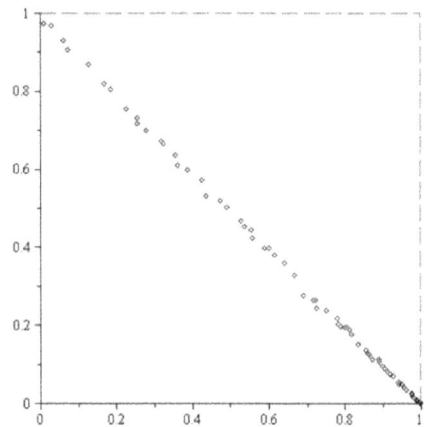

Figure 115: Pareto frontier highlighted for the example problem.

Monte Carlo Method

I-1	Independent variables $\vec{x}$

Objective function

$$\vec{F}(\vec{x}) = \begin{bmatrix} f_1(x) \\ f_2(x) \\ \vdots \\ f_n(x) \end{bmatrix}$$

I-2

I-3   Method for selecting points $\vec{x}$ randomly from the parameter space

S-1   Chose a point $\vec{x}$ from the parameter space

S-2   Compute the value of the objective function $\vec{F}(\vec{x})$

S-3   Compare the value of $\vec{F}(\vec{x})$ with each $\vec{F}(\vec{p}_i)$ on the list of points on the frontier.

       i.   If $\vec{F}(\vec{x})$ dominates $\vec{F}(\vec{p}_i)$, remove $\vec{F}(\vec{p}_i)$ from the list of frontier points. Continue checking the frontier points.

       ii.   If $\vec{F}(\vec{x})$ is dominated by $\vec{F}(\vec{p}_i)$, discard $\vec{F}(\vec{x})$ and continue from S-1.

S-4   If none of the points on the current list of frontier points dominates $\vec{F}(\vec{x})$, add $\vec{F}(\vec{x})$ to the list of frontier points.

S-5   Continue selecting points from S-1.

O-1   The list of frontier points $\vec{F}(\vec{p}_i)$ is the approximation to the Pareto frontier.

Algorithm XXXIII: The Monte Carlo method may be used to identify the Pareto frontier.

8.4 Weighted Sum Method

The weighted sum method constructs a series of single objective functions from the components of the multiobjective function, and then optimizes each of the single objective functions. The optimum of the single objective functions is a candidate for the Pareto frontier.

Let the multiobjective function be given by

$$\vec{F}(\vec{x}) = \begin{bmatrix} f_1(\vec{x}) \\ f_2(\vec{x}) \\ \vdots \\ f_n(\vec{x}) \end{bmatrix} \qquad 8.8$$

Based on this, consider the single objective

$$g(\vec{x}; \vec{\alpha}) = \sum_{i=1}^{n} \alpha_i f_i(\vec{x}) \qquad 8.9$$

where

$$\sum_{i=1}^{n} \alpha_i = 1 \qquad 8.10$$

The coefficients α_i may be viewed as the vector $\vec{\alpha}$. The algorithm proceeds by selecting particular values for $\vec{\alpha}$, then maximizing $g(\vec{x}; \vec{\alpha})$. The value of $\vec{x}$ that maximizes $g(\vec{x}; \vec{\alpha})$ represents a particular value as a candidate for a member of the Pareto frontier $\vec{F}(\vec{x})$.

Figure 116 provides a visual representation of the weighted sum method for a biobjective function. In this case the function under maximization has the form

$$\vec{F}(\vec{x}) = \begin{bmatrix} f_1(\vec{x}) \\ f_2(\vec{x}) \end{bmatrix} \qquad 8.11$$

The function $g(\vec{x}; \vec{\alpha})$ is

$$g(\vec{x}; \alpha) = \alpha f_1(\vec{x}) + (1 - \alpha) f_2(\vec{x}) \qquad 8.12$$

To implement the method, we select a particular value for α, then find the value of $\vec{x}$ that maximizes $g(\vec{x}; \alpha)$. Designate the value of $\vec{x}$ that maximizes $g(\vec{x}; \alpha)$ as $\vec{x}_i^*$. The value of $\vec{x}_i^*$ is then substituted into $\vec{F}(\vec{x})$ and is a candidate for the Pareto frontier. Designate $\vec{F}(\vec{x}_i^*)$ as $\vec{F}_i^*$. This is the ith candidate for a point on the Pareto frontier.

This process is repeated for different values of α. For each value of α, the function $g(\vec{x}; \alpha)$ is maximized to find $\vec{x}_i^*$. From this we compute $\vec{F}_i^*$, and create a list of candidates for the frontier.

To understand the methodology behind this, consider the particular case when $\alpha = 1$. In this case $g(\vec{x}; \alpha)$ is simply

$$g(\vec{x}; \alpha) = f_1(\vec{x}) \qquad 8.13$$

Maximizing $g(\vec{x}; \alpha)$ is the same as finding the global maximum for $f_1(\vec{x})$. In Figure 116, the maximum for $f_1(\vec{x})$ is the point whose projection onto the horizontal axis is the farthest from the origin. This is indicated by the vertical line extending from the horizontal axis to the operational space in Figure 116.

Alternatively, when $\alpha = 1$ we have

$$g(\vec{x}; \alpha) = f_2(\vec{x}) \qquad 8.14$$

In this case, maximizing $g(\vec{x}; \alpha)$ is the same as maximizing $f_2(\vec{x})$. The global maximum for $f_2(\vec{x})$ is the highest vertical point in Figure 116.

Continuing this process, when $\alpha = .5$ we have

$$g(\vec{x}; \alpha) = \frac{f_1(\vec{x}) + f_2(\vec{x})}{2} \qquad 8.15$$

To find the maximum, we draw the line in the operational space along the vector $\langle .5, .5 \rangle$ (the coefficients in 8.15) and project the operational region onto the vector. The point in operational space that has the largest projection onto the vector is the maximizer $\vec{x}_i^*$.

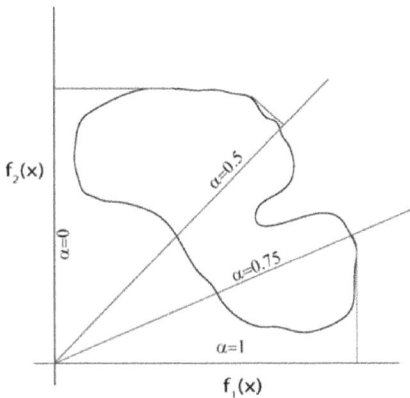

Figure 116: Weighted sum method with a biobjective function.

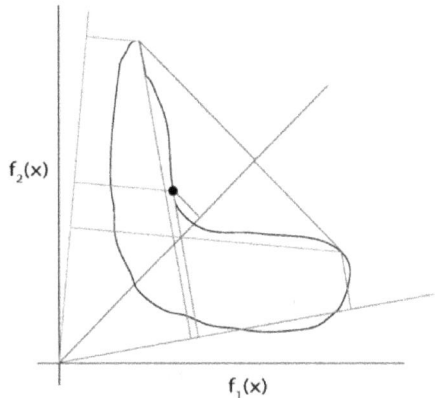

Figure 117: Example point in a concave region. No choice of weights can identify the point.

The weighted sum method is not able to identify every point on the frontier. Figure 117 provides an example of a point in a biobjective maximization problem that is not identifiable using the weighted sum method. A point in a concave region cannot be the maximum of the function $g(\vec{x}; \alpha)$ for any choice of α. For every line defined by a choice of α, there is another point in the optimization space whose projection is higher than the point in the concave region.

For example, Figure 117 shows three choices for α and the projections for three different points: a point in the concave region, a point at the upper extreme and a point at the left extreme. When the value of α is near zero, the line is near vertical and the projection from the left extreme is low. However, the projection from the upper extreme is higher than the projection from the concave point. Thus, optimizing $g(\vec{x}; \alpha)$ results in a point near the upper extreme.

Alternatively, choosing α near one results in a near horizontal line. In this case, the projection from the upper extreme is low, but the projection from the left extreme is high. In this case optimizing $g(\vec{x}; \alpha)$ results in a point near the left extreme away from the concave point.

Finally, choosing a value for α in between the extremes and maximizing $g(\vec{x}; \alpha)$ still results in a value near one of the extremes away from the concave point. Since the point is in the concave region, the points in the extremes project higher than the point in the concave region. Again, the point in the concave region does not maximize $g(\vec{x}; \alpha)$ and as a bresult, is not identified by the weighted sum method.

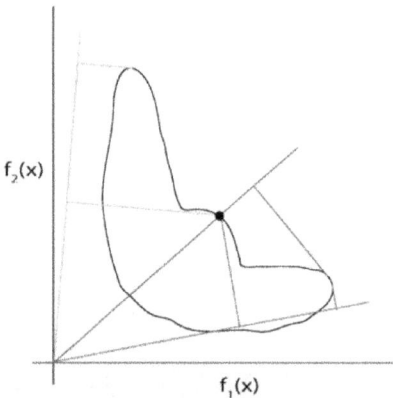

Figure 118: Frontier points in convex regions may not be identifiable using the weighted sum method.

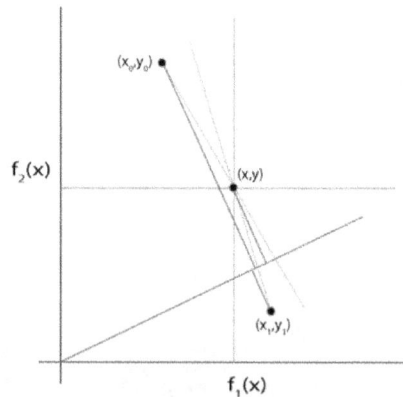

Figure 119: Point under consideration compared to the extreme slopes in the incomparable regions.

In fact, even points that are on convex regions of the frontier may not be identifiable due to the global behaviors of the frontier. Figure 118 shows an example of a point in a convex region of the frontier that is not identifiable using the weighted sum method. Although the point is in a locally convex region, the upper and left extremes put the point in a globally concave region.

The above arguments may be quantified to determine when a particular point may be identifiable using the weighted sum method. Figure 119 provides a test point located at (x, y) that is assumed to be on the frontier. The question we examine is under what conditions do other points make it impossible for the point at (x, y) to maximize the function $g(\vec{x}; \alpha)$. If there is no value of α where $g(\vec{x}; \alpha)$ is maximized by the point (x, y), then (x, y) will never be recognized by the weighted sum method.

There are four regions in relation to the point (x, y): points to the upper right, upper left, lower left, and lower right. There cannot be any points in the operational region to the upper right because these points would dominate (x, y) and hence the point could not be on the frontier. Thus, we do not need to consider this region further. In addition, any point to the lower left is dominated by the point (x, y). Points in this region never cause issue because if we choose the value of α that results in a line from the origin to the point (x, y), the point (x, y) always has a higher projection onto this line than any point to the lower left of (x, y). Thus, no point in this region ever prevents the method from recognizing (x, y).

The points to the upper left and lower right can prevent the method from recognizing (x, y). Examine a point to the upper left of (x, y). In Figure 119, the point (x_0, y_0) is a point in this region. A particular value of α creates a line where each operational point is projected. The projection of the point (x, y) to this line is

$$\alpha x + (1 - \alpha)y \qquad \text{8.16}$$

The projection of the point (x_0, y_0) is similarly

$$\alpha x_0 + (1 - \alpha)y_0 \qquad \text{8.17}$$

In order for the projection of (x, y) to be larger than the projection of (x_0, y_0) we need

$$\alpha x + (1 - \alpha)y > \alpha x_0 + (1 - \alpha)y_0 \qquad \text{8.18}$$

or

$$\alpha\Delta x + (1 - \alpha)\Delta y > 0 \qquad\qquad 8.19$$

where $\Delta y = y - y_0$ and $\Delta x = x - x_0$. Solving for α,

$$\alpha > -\frac{\Delta y}{\Delta x - \Delta y} \qquad\qquad 8.20$$

In terms of the slope of the line joining the points $(s_0 = \Delta y / \Delta x)$,

$$\alpha > -\frac{s_0}{1 - s_0} \qquad\qquad 8.21$$

The slope of the line is always negative, so the above expression is on the range $\alpha \in (0,1)$. Similarly for the point (x_1, y_1) in the lower right region we have

$$\alpha < -\frac{s_1}{1 - s_1} \qquad\qquad 8.22$$

Again, since the slope is always negative, α is on the range $(0,1)$.

Each of these expressions takes the form

$$\frac{x}{1 + x} \qquad\qquad 8.23$$

where x is a positive value. Figure 120 provides a graph of this function.

These two expressions place boundaries on the possible values of α that result in (x, y) maximizing $g(\vec{x}; \alpha)$. Consider an arbitrary point (x, y) on the Pareto frontier. Based on this point, find the point on the frontier in the upper left region that results in the maximum magnitude for slope of the line joining (x, y). Designate this slope as s_u. Similarly, find the point on the frontier to the lower right that has the minimum magnitude for the slope s_l.

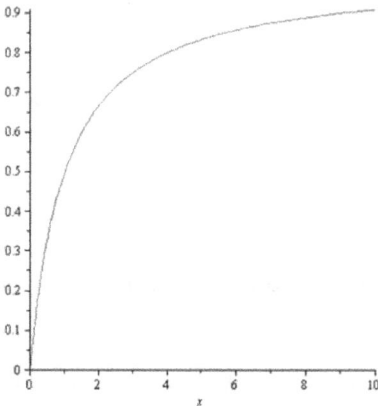

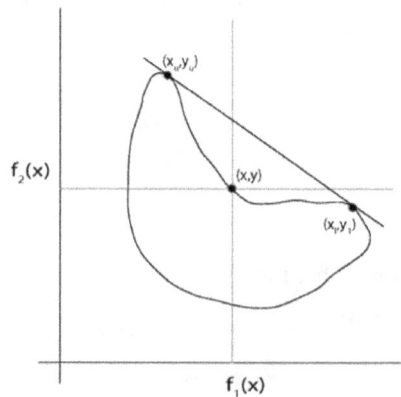

Figure 120: Plot of the slope constraint for the weighted sum method.

Figure 121: If 8.24 is not met, the point is not obtainable from the weighted sum method.

If

$$|s_u| < |s_l|$$ 8.24

then there exist a value of α such that (x, y) maximizes $g(\vec{x}; \alpha)$. Otherwise the point (x, y) is not obtainable using the weighted sum method.

Weighted Sum Method

I-1   Independent variables $\vec{x}$

Objective function

I-2
$$\vec{F}(\vec{x}) = \begin{bmatrix} f_1(\vec{x}) \\ f_2(\vec{x}) \\ \vdots \\ f_n(\vec{x}) \end{bmatrix}$$

I-3   Method for optimizing a single objective function $g(\vec{x})$

Chose a point $\vec{\alpha}$ such that

S-1
$$\sum_{i=1}^{n} \alpha_i = 1 \quad |\alpha_i| \le 1$$

Maximize the single objective function

S-2
$$g(\vec{x}) = \sum_{i=1}^{n} \alpha_i f_i(\vec{x})$$

Let $\vec{x}^*$ designate this maximum.

S-3   Compute the value of $\vec{F}(\vec{x}^*)$

S-4   Add $\vec{F}(\vec{x}^*)$ to the list of potential points on the Pareto frontier and continue from S-1 for the desired number of $\vec{\alpha}$ to compute

S-5   Iterate through the list of potential points for the Pareto frontier

S-6   Consider the ith point on the list $\vec{F}(\vec{x}_i^*)$

S-7   Compare $\vec{F}(\vec{x}_i^*)$ to each other point on the list of potential frontier points. If $\vec{F}(\vec{x}_i^*)$ is dominated by another point $\vec{F}(\vec{x}_j^*)$, eliminate $\vec{F}(\vec{x}_i^*)$ from the list. If $\vec{F}(\vec{x}_i^*)$ dominates any other point $\vec{F}(\vec{x}_k^*)$, eliminate $\vec{F}(\vec{x}_k^*)$ from the list.

O-1   The list of frontier points $\vec{F}(\vec{x}_i^*)$ remaining after S-7 is the approximation to the Pareto frontier.

Algorithm XXXIV: The weighted sum method may be used to identify the Pareto frontier.

8.5 Normal-Boundary Intersection

The normal boundary intersection method transforms an unconstrained multiobjective optimization problem to a series of multiply constrained single objective optimization problem. Thus, the single objective optimization techniques developed earlier may be employed to solve for the Pareto frontier.

Let the optimization function have the form

$$\vec{F}(\vec{x}) = \begin{bmatrix} f_1(\vec{x}) \\ f_2(\vec{x}) \\ \vdots \\ f_n(\vec{x}) \end{bmatrix} \qquad 8.25$$

First, find the global optimum for each component of the objective function. Designate the global optimum for $f_i(\vec{x})$ as $\vec{x}_i^*$. Furthermore, as a notational convenience let $\vec{F}(\vec{x}_i^*) = \vec{F}_i^*$. Finally, let

$$\vec{F}^* = \begin{bmatrix} f_1(\vec{x}_1^*) \\ f_2(\vec{x}_2^*) \\ \vdots \\ f_n(\vec{x}_n^*) \end{bmatrix} \qquad 8.26$$

The last expression is the vector whose components are each set to their respective global optimum and is called the utopia point. It is unlikely that this vector is physically realizable as this only occurs when the Pareto frontier is a single point that dominates all other points in the operational space. Figure 122 provides examples of the points $\vec{F}_1^*$, $\vec{F}_2^*$, and $\vec{F}^*$ in a simple biobjective case.

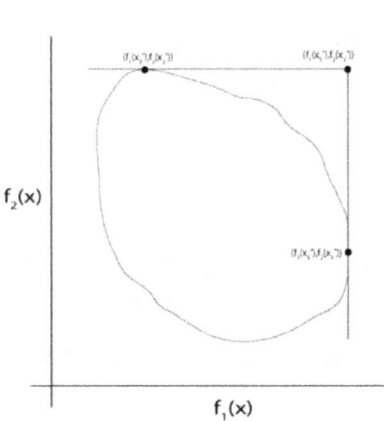

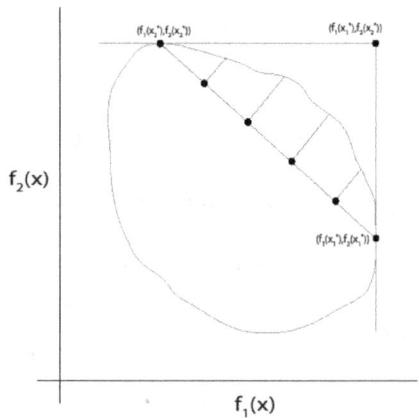

Figure 122: Objective space showing the global maximums for each of the components and the utopia point.

Figure 123: Sample points on the optimal hyperplane in the normal-boundary intersection method.

Consider the matrix

$$\Phi = \begin{bmatrix} 0 & f_2(\vec{x}_2^*) - f_1(\vec{x}_2^*) & \cdots & f_n(\vec{x}_n^*) - f_1(\vec{x}_n^*) \\ f_1(\vec{x}_1^*) - f_2(\vec{x}_1^*) & 0 & \cdots & f_n(\vec{x}_n^*) - f_2(\vec{x}_n^*) \\ \vdots & \vdots & \vdots & \vdots \\ f_1(\vec{x}_1^*) - f_n(\vec{x}_1^*) & f_2(\vec{x}_2^*) - f_n(\vec{x}_2^*) & \cdots & 0 \end{bmatrix} \quad 8.27$$

This is the matrix whose ith column is the vector $\vec{F}^* - \vec{F}_i^*$. Let $\vec{w}$ be a weight vector such that $|w_i| \le 1$ and

$$\sum_{i=1}^{n} w_i = 1 \quad\quad 8.28$$

Then $\Phi\vec{w}$ is a vector in the hyperplane defined from the $\vec{F}^* - \vec{F}_i^*$ vectors. This is the hyperplane that passes through the global optimum for each individual component. Let $\hat{n}$ be a unit normal vector perpendicular to this hyperplane and directed away from the origin. If we select a set of values for $\vec{w}$, then the vector $\Phi\vec{w}$ parallel to $\hat{n}$ must intersect the operational region at some point at the boundary of the operational region. This point of intersection is a candidate for a point on the Pareto frontier.

A candidate point is computed by finding the largest value of t such that there exists some value of $\vec{x}$ where

$$\Phi\vec{w} + t\hat{n} = \vec{F}(\vec{x}) \quad\quad 8.29$$

Equation 8.29 is a vector relation. This is a set of n constraints that must be simultaneously met.

In practice the vector normal to the hyperplane of optimal values may be difficult to compute. However, the method works with any vector pointed toward the Pareto frontier. In many cases, it may be easier to use the vector

$$\Phi\vec{u} \quad\quad 8.30$$

where $\vec{u}$ is the vector with all ones as components: $\vec{u} = \langle 1, 1, \dots, 1 \rangle$. Using this vector instead of the normal vector the conditions 8.29 become

$$\Phi[\vec{w} + t\vec{u}] = \vec{F}(\vec{x}) \quad\quad 8.31$$

The normal boundary intersection method changes an unconstrained multiobjective optimization problem into a series of constrained single objective optimization problems. These single objective optimization problems attempt to maximize the value t subject to the constraints 8.29. There are n constraints in 8.29, one for each component in the vector.

Figure 123 shows an example of implementing the normal-boundary intersection method in a biobjective case. First we identify the global maximum for each component of the optimization space. Next, we construct the utopia line which is the line connecting the optimal vertical point with the optimal horizontal point. Then we select a set of weight points along the utopia line. In the example, the weights are selected at equal intervals along the line. Finally, we optimize the distance along the normal line beginning on the utopia line. The optimal value provides a point on the boundary of the operational space.

For two dimensional objectives, every point on the Pareto frontier is a solution to some problem of the form 8.31. However, this is not true in higher dimensions. Thus, in two dimensions the normal-boundary intersection method is capable of identifying the entire frontier, but in higher dimensions there may exist points on the frontier that the method is not able to recognize.

As an example of the implementation of the normal-boundary intersection method, consider the multiobjective optimization problem

$$\vec{F}(r,s) = \begin{bmatrix} r\cos\left(\frac{\pi}{2}s\right) \\ r\sin\left(\frac{\pi}{2}s\right) \end{bmatrix} \qquad\qquad 8.32$$

where $r, s \in [0,1]$. In this optimization we desire to maximize the objectives.

First we find the global maximum for each component of the objective. The maximum for first component is at the point $(r,s) = (1,0)$, and the maximum for the second component is $(r,s) = (0,1)$.

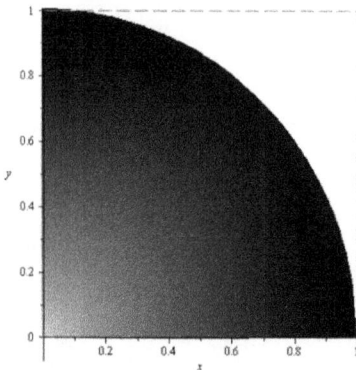

Figure 124: Sample multiobjective space. Brighter areas have a higher density from the parameter space.

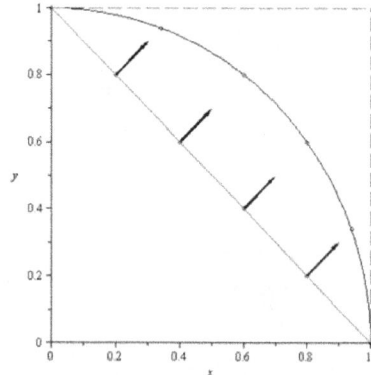

Figure 125: Estimate to the Pareto frontier using four sample points under the normal-boundary intersection method.

The objective space is shown in Figure 124. The objective space is the upper-right quadrant unit semicircle. The maximum value for the x-component is the value one and occurs when the y-component is at its minimum value of zero. Similarly, the maximum value for the y-component is also one and occurs when the x-component is at its minimum value of zero.

The utopia point is the point $(1,1)$. This point is composed of the maximum value for the x-axis (1) and the maximum value of the y-axis (1). This point is not achievable because it is not in the operational space.

Figure 125 Shows the Pareto frontier along with the utopia point. In terms of components

$$\vec{x}_1^* = (1,0) \tag{8.33}$$

$$\vec{x}_2^* = (1,1) \tag{8.34}$$

$$f_1(\vec{x}_1^*) = 1 \tag{8.35}$$

$$f_2(\vec{x}_1^*) = 0 \tag{8.36}$$

$$f_1(\vec{x}_2^*) = 0 \tag{8.37}$$

$$f_2(\vec{x}_2^*) = 1 \tag{8.38}$$

$$\vec{F}_1^* = \begin{bmatrix} 1 \\ 0 \end{bmatrix} \tag{8.39}$$

$$\vec{F}_2^* = \begin{bmatrix} 0 \\ 1 \end{bmatrix} \tag{8.40}$$

$$\vec{F}^* = \begin{bmatrix} 1 \\ 1 \end{bmatrix} \tag{8.41}$$

From these results, the matrix Φ is

$$\Phi = \begin{bmatrix} 0 & f_2(\vec{x}_2^*) - f_1(\vec{x}_2^*) \\ f_1(\vec{x}_1^*) - f_2(\vec{x}_1^*) & 0 \end{bmatrix} \tag{8.42}$$

$$= \begin{bmatrix} 0 & 1-0 \\ 1-0 & 0 \end{bmatrix} \tag{8.43}$$

$$= \begin{bmatrix} 0 & 1 \\ 1 & 0 \end{bmatrix} \tag{8.44}$$

The utopia line is the line between the points $(1,0)$ and $(0,1)$ in the operational space. These corresponding points in the parameter are $(1,0)$ and $(1,1)$.

We can use either equation 8.29 or 8.31 as the constrain equation. Under 8.29, we need to compute the vector normal to the line connecting the points $(1,0)$ and $(0,1)$. The normal vector may be found from the vector difference of these points:

$$\vec{d} = \langle 1,0 \rangle - \langle 0,1 \rangle = \langle 1,-1 \rangle \qquad 8.45$$

There are two vectors perpendicular to this vector:

$$\vec{p}_1 = \langle 1,1 \rangle \qquad 8.46$$

$$\vec{p}_2 = \langle -1,-1 \rangle \qquad 8.47$$

The first vector points toward the utopia point, while the second vector points toward the origin. Since we desire to maximize the objectives, we use the first vector pointing toward the utopia point. This vector is in the direction of the normal vector. The normal may be computed from this by dividing by the magnitude:

$$\hat{n} = \frac{1}{\sqrt{2}} \langle 1,1 \rangle \qquad 8.48$$

If we use the conditions from 8.31, the vector $\vec{u}$ is $\langle 1,1 \rangle$. In this case, the normal vector and $\vec{u}$ point in the same direction. Essentially, these result in the same set of conditions and only scale the parameter t relative to each other. We continue using the condition 8.31:

$$\begin{bmatrix} 0 & 1 \\ 1 & 0 \end{bmatrix} \begin{bmatrix} w+t \\ 1-w+t \end{bmatrix} = \begin{bmatrix} r\cos\left(\frac{\pi}{2}s\right) \\ r\sin\left(\frac{\pi}{2}s\right) \end{bmatrix} \qquad 8.49$$

$$\begin{bmatrix} 1-w+t \\ w+t \end{bmatrix} = \begin{bmatrix} r\cos\left(\frac{\pi}{2}s\right) \\ r\sin\left(\frac{\pi}{2}s\right) \end{bmatrix} \qquad 8.50$$

We seek to optimize the value t subject to these constraints. This is a single objective optimization problem subject to two constraints. The constraints may be incorporated into a single objective function by taking the absolute value of difference of each constraint:

$$L(t,s,t) = t - \left|1-w+t-r\cos\left(\frac{\pi}{2}s\right)\right| - \left|w+t-r\sin\left(\frac{\pi}{2}s\right)\right| \qquad 8.51$$

Choose four values to sample w: .2, .4, .6, .8. At each of these values of w, the single objective function 8.51 is maximized. In this sample, the amoeba method from section 2.24 is used to find the optimal value of t and the corresponding values of r, s. The results of these four optimizations are shown in Figure 125.

Normal-Boundary Intersection

I-1   Independent variables $\vec{x}$

     Objective function

I-2
$$\vec{F}(\vec{x}) = \begin{bmatrix} f_1(\vec{x}) \\ f_2(\vec{x}) \\ \vdots \\ f_n(\vec{x}) \end{bmatrix}$$

I-3   Method for optimizing a single unconstrained objective function

I-4   Method for optimizing the single objective function $g(t) = t$ under a set of n constraints of the form $\Phi[\vec{w} + t\vec{u}] = \vec{F}(\vec{x})$

S-1   Compute the global optimum for each individual component of the objective function. Let $\vec{x}_i^*$ designate the global optimum for $f_i(\vec{x})$

     Compute the matrix

S-2
$$\Phi = \begin{bmatrix} 0 & f_2(\vec{x}_2^*) - f_1(\vec{x}_2^*) & \cdots & f_n(\vec{x}_n^*) - f_1(\vec{x}_n^*) \\ f_1(\vec{x}_1^*) - f_2(\vec{x}_1^*) & 0 & \cdots & f_n(\vec{x}_n^*) - f_2(\vec{x}_n^*) \\ \vdots & \vdots & \vdots & \vdots \\ f_1(\vec{x}_1^*) - f_n(\vec{x}_1^*) & f_2(\vec{x}_2^*) - f_n(\vec{x}_2^*) & \cdots & 0 \end{bmatrix}$$

     Choose a weight vector $\vec{w}$ such that

S-3
$$\sum_{i=1}^{n} w_i = 1 \quad |w_i| \leq 1$$

     Find the largest value t such that the n constraints

S-4
$$\Phi[\vec{w} + t\vec{u}] = \vec{F}(\vec{x})$$

     are satisfied. Let $\vec{x}_i^b$ be the value of $\vec{x}$ corresponding to the maximum.

S-5   Save the point $\vec{F}(\vec{x}_i^b)$ as a candidate for the Pareto frontier.

S-6   Repeat from S-3 for as many points as desired.

S-7   Compare $\vec{F}(\vec{x}_i^b)$ to each other point on the list of potential frontier points. If $\vec{F}(\vec{x}_i^b)$ is dominated by another point $\vec{F}(\vec{x}_j^b)$, eliminate $\vec{F}(\vec{x}_i^b)$ from the list. If $\vec{F}(\vec{x}_i^b)$ dominates any other point $\vec{F}(\vec{x}_K^b)$, eliminate $\vec{F}(\vec{x}_i^b)$ from the list.

O-1   The list of frontier points $\vec{F}(\vec{x}_i^b)$ remaining after S-7 is the approximation to the Pareto frontier.

Algorithm XXXV: The normal-boundary method may be used to identify the Pareto frontier.

8.6 Normal Constraint

The normal constraint method shares many of the same features as the normal-boundary intersection method. Both methods use the utopia hyperplane to exchange an unconstrained multiobjective optimization problem into a multiply constrained single objective optimization problem.

Let the optimization function have the form

$$\vec{F}(\vec{x}) = \begin{bmatrix} f_1(\vec{x}) \\ f_2(\vec{x}) \\ \vdots \\ f_n(\vec{x}) \end{bmatrix} \qquad 8.52$$

The normal constraint method begins by computing the global optimum for each component of the objective function individually. Let $\vec{x}_i^*$ be the optimum parameter vector for the component $f_i(\vec{x})$, and set $\vec{F}(\vec{x}_i^*) = \vec{F}_i^*$. The utopia point is the vector whose components are the optimum values for each component individually:

$$\vec{F}^* = \begin{bmatrix} f_1(\vec{x}_1^*) \\ f_2(\vec{x}_2^*) \\ \vdots \\ f_n(\vec{x}_n^*) \end{bmatrix} \qquad 8.53$$

The utopia point is not typically within the multiobjective operational space. If the utopia point is in the operational space, then this point is the unique optimum as this point dominates all other feasible points.

The utopia vectors are the n vectors where the ith component has the value $f_n(\vec{x}_n^*)$ and all other components are zero:

$$\vec{U}^i = \begin{bmatrix} 0 \\ 0 \\ \vdots \\ f_i(\vec{x}_i^*) \\ \vdots \\ 0 \end{bmatrix} \qquad 8.54$$

The utopia point is the sum of the utopia vectors

$$\vec{F}^* = \sum_{i=1}^{n} \vec{U}^i \qquad 8.55$$

Figure 126 provides an example of an operational space where the maximum values for the components are considered the optimums, and the utopia point is constructed from these optima.

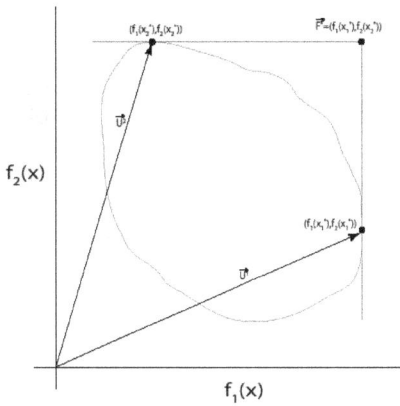

Figure 126: Objective space showing the global maximums for each of the components and the utopia point and the utopia vectors.

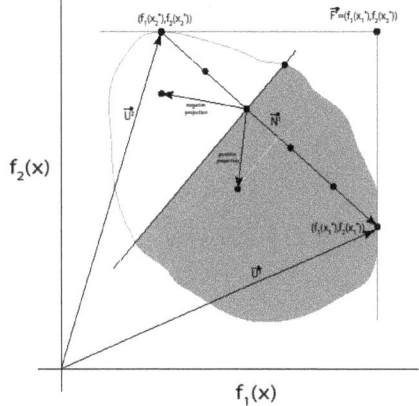

Figure 127: Implementation of the normal constraint method. The shaded region is the region that meets the constraint.

The next step is to 'normalize' the objective function. This is one of the key elements of the normal constraint method. By normalizing the objective function, the normal constraint method is independent of the scale of the problem. This means that if we were to independently scale each component, it would not make any difference to the result of the optimization. This scale independence is a desirable feature for multiobjective optimization methods that identify the Pareto frontier.

To compute the normalization, we need to compute the Nadir point. The Nadir point is a vector where each component is minimum value for that component across the set of optimum parameter vectors:

$$\vec{N} = \langle N_1, N_2, \dots, N_n \rangle \qquad 8.56$$

where

$$N_i = min[f_i(\vec{x}_1^*), f_i(\vec{x}_2^*), \dots, f_i(\vec{x}_n^*)] \qquad 8.57$$

The normalization process transforms each component of the objective function independently. The ith component of the objective function is transformed as

$$\bar{f}_i(\vec{x}) = \frac{f_i(\vec{x}) - N_i}{f_i(\vec{x}_i^*) - N_i} \qquad 8.58$$

Under this transformation, if we evaluate each component over the set of optimum parameter vectors $\vec{x}_n^*$, the transformed components $\bar{f}_i(\vec{x})$ are on the range $[0,1]$. Since we have transformed the components using properties of the objective, the transformation is independent of the scaling of the components.

The transformed objective function is

$$\vec{F}(\vec{x}) = \begin{bmatrix} \bar{f}_1(\vec{x}) \\ \bar{f}_2(\vec{x}) \\ \vdots \\ \bar{f}_n(\vec{x}) \end{bmatrix} \qquad 8.59$$

and the transformed utopia vectors $\vec{U}^i$ are each unit vectors. The utopia point in the transformed coordinates is the vector $\langle 1,1, \ldots, 1 \rangle$.

The utopia hyperplane may be defined in terms of a set of vectors spanning the hyperplane. One such set of vectors is

$$\vec{H}^k = \vec{U}^k - \vec{U}^n \qquad 8.60$$

This is a set of $n-1$ vectors ($\vec{H}^n$ is the null vector) that span the utopia hyperplane. Using the hyperplane vectors, choose a set of equally distributed points in the hyperplane as initial points for the algorithm. Equally distributed points in the hyperplane may be found as

$$\vec{X}^j = \sum_{k=1}^{n} \alpha_{kj} \vec{U}^k \qquad 8.61$$

where

$$\sum_{k=1}^{n} \alpha_{kj} = 1 \quad |\alpha_{kj}| \leq 1 \qquad 8.62$$

The 'constraint' part of the normal constraint method divides the operational space into allowable and non-allowable regions based on the utopia hyperplane and a seed point $\vec{X}^j$. The allowable region is the set of points in the operational space such that

$$\vec{H}^k \cdot \left(\vec{F}(\vec{x}) - \vec{X}^j \right) \geq 0 \qquad 8.63$$

There are $n-1$ such constraints, one for each $\vec{H}^k$. Figure 127 provides an example of the allowable and non-allowable regions for a sample point $\vec{X}^j$. In this example, the utopia hyperplane is the line segment joining the points at the maximum values for each of the components. There are four test points $\vec{X}^j$ along the utopia line (hyperplane). The diagram provides an example of the method at the second point.

The constraints 8.63 examine the dot product of two vectors. The vector on the right is the vector pointing from $\vec{X}^j$ to an arbitrary point $\vec{F}(\vec{x})$ in the objective space. The vector $\vec{H}^k$ is the vector pointing from $\vec{U}^n$ to the maximum value for the kth component ($\vec{U}^k$). All of the vectors $\vec{H}^k$ point away from $\vec{U}^n$.

The projection of the operational point onto $\vec{H}^k$ is either positive or negative depending on the location of the operational point. If the projection is negative, then the point lies in the infeasible region and it is rejected. If the projection is positive, then the point lies in the feasible region and is accepted.

Examining the feasible region (points meeting the constraints 8.63), the maximum value of $\bar{f}_n(\vec{x})$ corresponds to a point on the boundary of the feasible space. Thus, if we maximize the single objective $\bar{f}_n(\vec{x})$ subject to the constraints in 8.63, the result is a point on the boundary of the operational space that is a candidate for a point on the Pareto frontier.

One advantage that the normal constraint method has over the normal-boundary intersection method is that the normal constraint method is able to filter some points that are not on the frontier. Figure 128 provides an example comparing the normal constraint method to the normal-boundary intersection method. Both use the utopia hyperplane as the basis for creating a single objective optimization. The normal-boundary intersection method identifies the point on the boundary at NBI. This point is not on the frontier and is dominated by the point NC identified by the normal constraint method.

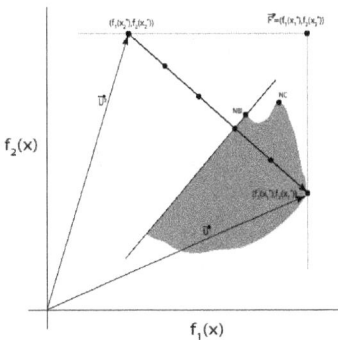

Figure 128: Normal constraint (NC) method compared with the normal-boundary intersection (NBI) method for a sample point on the utopia hyperplane. The NC method identifies the point NC whereas the NBI method produces the point at NBI. The point NC is on the Pareto frontier and dominates the point NBI.

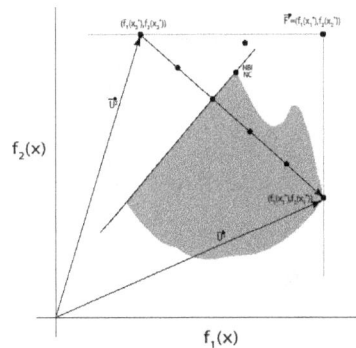

Figure 129: Another example comparing the normal constraint (NC) method with the normal-boundary intersection (NBI) method. In this case, both NC and NBI produce the same point. This point is dominated by another point on the frontier. IN this situation the NC method identifies boundary points that are not in fact on the frontier.

However, the normal constraint method does identify points on the boundary that are not on the frontier. Figure 129 provides an example comparing the normal constraint and normal-boundary methods at a different point under the same objective as in Figure 128. In this case, both the normal constraint and normal-boundary intersection methods identify the same point. This point is in fact dominated by another point on the frontier as indicated in the figure.

As an example problem, consider the operational space defined by

$$f_1(x, y) = xy \qquad\qquad 8.64$$

$$f_2(x, y) = \frac{y}{1.0015}\left[e^{-10(x-.25)^2} + .8e^{-25(x-.75)^2}\right] \qquad 8.65$$

The operational space defined by these component equations is shown as the darker circles in Figure 130 and Figure 131. The operational space is bimodal with a segment of the frontier dominated between the peaks.

We apply both the normal-boundary intersection and normal constraint method to this objective. Figure 130 shows the results of the normal-boundary intersection method. The lighter circles indicate the frontier points identified by the method. We see that the method adheres well to the frontier and identifies the main features.

Alternatively, Figure 131 shows the application of the normal constraint method to the same objective. Again, the method adheres well to the frontier. In fact, both methods identify nearly identical points on the frontier.

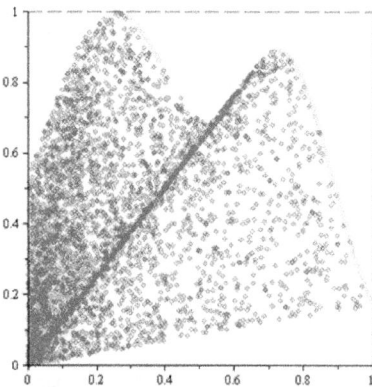

Figure 130: Normal-boundary intersection method applied to the example problem. Darker circles are samples of the operational space while lighter circles are the frontier points discovered by NBI.

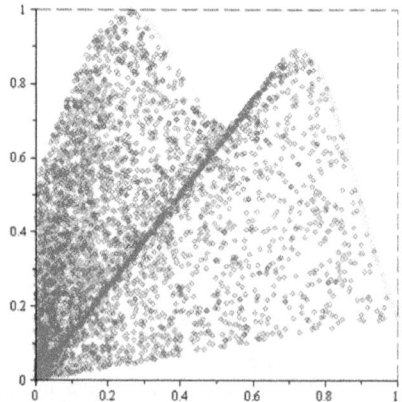

Figure 131: Normal constraint method applied to the example problem. Darker circles are samples of the operational space while lighter circles are the frontier points discovered by NC.

Normal Constraint

I-1 Independent variables $\vec{x}$

I-2 Objective function $\vec{F}(\vec{x})$ with components $f_n(\vec{x})$

I-3 Method for optimizing a single unconstrained objective function

I-4 Method for optimizing the single objective function $f_n(\vec{x})$ under a set of n constraints of the form $\vec{\vec{H}}^k \cdot \left(\vec{F}(\vec{x}) - \vec{\vec{X}}^j\right) \geq 0$

S-1 Compute the global optimum for each individual component of the objective function. Let $\vec{x}_i^*$ designate the global optimum for $f_i(\vec{x})$

S-2 Compute the Nadir point $\vec{N} = \langle N_1, N_2, \dots, N_n \rangle$ where
$$N_i = min[f_i(\vec{x}_1^*), f_i(\vec{x}_2^*), \dots, f_i(\vec{x}_n^*)]$$

S-3 Transform the coordinates using
$$\bar{f}_i(\vec{x}) = \frac{f_i(\vec{x}) - N_i}{f_i(\vec{x}_i^*) - N_i}$$

S-4 Compute the $n-1$ hyperplane vectors that point from the point $(0,0,\dots,0,1)$ to each of the component maximums (these are unit vectors along each axis in the transformed coordinates)
$$\vec{\vec{H}}^k = \hat{e}^k - \hat{e}^n$$

S-5 Create a lattice of equally spaced points on the utopia hyperplane
$$\vec{\vec{X}}^j = \sum_{k=1}^{n} \alpha_{kj}\hat{e}^k \quad \sum_{k=1}^{n} \alpha_{kj} = 1 \quad |\alpha_{kj}| \leq 1$$

S-6 For each $\vec{\vec{X}}^i$, maximize $\bar{f}_n(\vec{x})$ subject to the constraints
$$\vec{\vec{H}}^k \cdot \left(\vec{F}(\vec{x}) - \vec{\vec{X}}^j\right) \geq 0$$
Let $\vec{x}_i^b$ be the corresponding point in parameter space.

S-7 Repeat from S-5 for each $\vec{\vec{X}}^i$.

S-8 Compare $\vec{F}(\vec{x}_i^b)$ to each other point on the list of potential frontier points. If $\vec{F}(\vec{x}_i^b)$ is dominated by another point $\vec{F}(\vec{x}_j^b)$, eliminate $\vec{F}(\vec{x}_i^b)$ from the list. If $\vec{F}(\vec{x}_i^b)$ dominates any other point $\vec{F}(\vec{x}_K^b)$, eliminate $\vec{F}(\vec{x}_i^b)$ from the list.

O-1 The list of frontier points $\vec{F}(\vec{x}_i^b)$ remaining after S-8 is the approximation to the Pareto frontier.

Algorithm XXXVI: The normal constraint method may be used to identify the Pareto frontier.

8.7 Strength Pareto Evolutionary Algorithm 2

The strength Pareto evolutionary algorithm 2 (SPEA2) is an evolutionary algorithm to find the Pareto frontier for a multiobjective optimization problem. SPEA2 defines two separate fitness measures for each state and sums these to arrive at the overall fitness.

The method utilizes traditional evolutionary algorithm principles to identify the Pareto frontier for a multiobjective optimization problem. To achieve this, the fitness for the evolutionary states is based in part on whether the state is dominated by other states.

To begin, a set of initial states is chosen randomly in the operating space. States are assigned to four different pools. A state may pass through all four pools during an iteration of the algorithm.

Population Pool – The population pool is comprised of states from the general population. This pool may have duplicate states. The population pool is initially filled with random states, but afterward, each iteration of the algorithm creates a new population pool from the mating pool. The population pool is combined with the archive set to create the union set.

Archive Set - The archive set is a set of nondominated states and is the approximation to the Pareto frontier for each iteration. The archive set only has unique members and does not have any duplicates. The archive set is created by identifying the nondominated states from the union set. The archive set is combined with the population pool to create the union set.

Union Set – The union set is created by combining the archive set with the population pool and retaining only unique states. The nondominated states of the union set make up the archive set for the next generation.

Mating Pool – The mating pool is created from the union set according to fitness. Binary tournament selection with replacement is used on the union set to create the mating pool. A pair of states are randomly selected from the union set, and the state with the lower fitness is placed into the mating pool. The process is repeated until the mating pool has the desired number of states. It is possible for the same state to be placed into the mating pool multiple times. Thus, the mating pool may contain duplicate states.

Initially, a set of random states is placed into the population pool, and the other pools are empty. The population pool and archive set are combined to create the union set. Although the population pool may have duplicate states, the union set contains only unique states.

A fitness is assigned to each state in the union set. The fitness has two components: dominance and density. We desire to identify a set of states that are on the Pareto frontier and are distributed over the entire hypersurface of the frontier. To achieve this, the fitness assigned to each state measures these values with smaller values more desirable.

The dominance component is computed in terms of the number of states each state dominates. First, for each individual in the union set, count the number of states dominated by the individual. Then, for each individual in the union state, sum the dominance count for each state that dominates the individual. Let p_j be the number of states dominated by state j. Furthermore, let '$i \prec j$' mean the set of states j that dominate state i. The dominance metric for state i is

$$S_i = \sum_{i \prec j} p_j \qquad\qquad 8.66$$

This means that the dominance metric for state i is computed by identifying all states j that dominate i, then summing the values of p_j for each of these states.

The dominance metric is higher for states that are dominated by many states. The lowest value for the dominance metric is zero which occurs if the state is not dominated by any other state.

The density metric is based on the distance between a given state and its kth nearest neighbor. For each state in the union set, compute the distance to every other state. The distance is the Euclidean distance between the state vectors in the operational space:

$$d\left(\vec{x}^i, \vec{x}^j\right) = \sqrt{\sum_{k=1}^{n} \left(x_k^i - x_k^j\right)^2} \qquad\qquad 8.67$$

Once the distances are computed, the resulting list is rank ordered from lowest to highest. The kth nearest neighbor is selected where k is typically chosen as

$$k = \sqrt{N + \bar{N}} \qquad\qquad 8.68$$

where N is the number of states in the population pool and $\bar{N}$ is the number of states in the archive.

The kth nearest neighbor is determined using the original states from the population pool and archive pool. Thus, when identifying the kth nearest neighbor, we account for the multiplicity of duplicate states in the population pool.

Once the kth nearest neighbor is identified, the density metric for state i is

$$D_i = \frac{1}{d_{ik} + 2} \qquad\qquad 8.69$$

where d_{ik} is the distance between the ith state and its kth nearest neighbor.

The total fitness for the ith state is given by the sum of the dominance metric and density metric:

$$F_i = S_i + D_i \qquad\qquad 8.70$$

Next, the archive set is created from the union set. Each element of the union set that is nondominated ($S_i = 0$) is placed into the next generation archive set. If there are exactly $\overline{N}$ states, the archive set is perfected and we can continue. Otherwise, we add or remove stated to the archive set so that there are exactly $\overline{N}$ states.

If states need to be added, examine the union set from the previous generation. Add the states from the previous union set that have the lowest fitness and are not already in the current archive. Thus, at first nondominated states are added because these states have $S_i = 0$ and necessarily have lower fitness than and dominated state. Once these are exhausted, the process continues by adding the best dominated states in order.

If states need to be removed, we identify the state with the smallest nearest neighbor distance (d_{i1}). This state is removed from the archive. This process is continued until the archive has the desired size. In the case where there are multiple states with the same value for d_{i1}, then we choose among these the state with the smallest d_{i2}. If there is still a tie, then examine d_{i3} and so on.

The mating pool is constructed from this archive set just created. This is created by applying binary tournament selection with replacement on the archive set. From the archive set, choose two states at random. Copy the state with the lower fitness into the mating pool, but do not remove either of these states from the archive set. Continue selecting pairs of states and copying the state with the lower fitness into the mating pool until the mating pool has the desired size.

Because the states are not removed from the archive as they are selected for the mating pool, it is possible that the mating pool contains duplicate states. This is a desired feature for the mating pool and effectively weights the mating pool in favor of states with superior fitness.

The next generation population pool is created by applying the traditional recombination and mutation algorithms to these states (see Genetic Algorithms §2.19). Once this population pool is created, the algorithm proceeds iteratively.

At each iteration, the archive set represents the approximation to the Pareto frontier.

As an example of the application of the SPEA2 method, consider the multiobjective

$$f_1(x, y) = \frac{x(1-y)}{2}(1 + sin(5\pi x)) \qquad 8.71$$

$$f_2(x, y) = \frac{y(1-x)}{2}(1 + cos(5\pi y)) \qquad 8.72$$

where $x, y \in [0,1]$. The operational space for this objective is shown in Figure 132. The objective has several sinusoidal bumps superimposed on a hyperbola.

Figure 132 shows the frontier identified using SPEA2 as dark circles, while Figure 133 shows the frontier identified by the method in isolation.

The SPEA2 algorithm identified 523 Pareto incomparable points after 1000 iterations. The algorithm is setup with a population pool size of 200, archive size of 100, and mating pool at 1000.

In this implementation, the SPEA2 algorithm is setup to construct the new population pool at each generation only using mutation. Recombination may be used but was not needed for the level of convergence shown in the figures.

Furthermore, all archive states are retained after each iteration and are combined together to create the nondominated states for the frontier. This way nondominated points are not thrown out each generation and additional frontier points are identified.

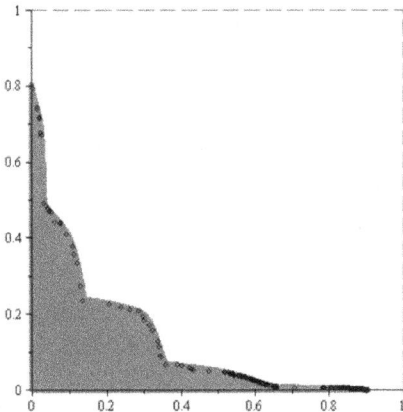

Figure 132: The SPEA2 method to an example problem. Lighter circles are samples of the operational space while darker circles are the frontier points discovered by SPEA2.

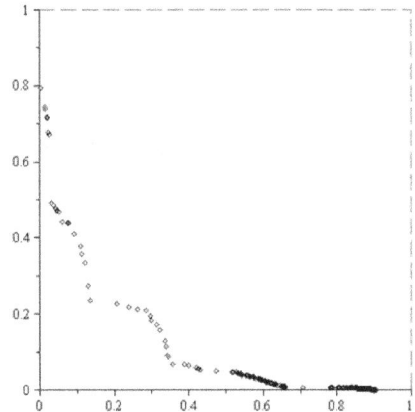

Figure 133: Pareto frontier identified by the SPEA2 algorithm.

SPEA2

I-1   Independent variables $\vec{x}$

I-2   Objective function $\vec{F}(\vec{x})$ with components $f_n(\vec{x})$

I-3   Evolutionary method for generating a new population based on the fitness values for the states in the current population.

I-4   Size of the population pool N

I-5   Size of the archive $\bar{N}$

I-6   Size of the mating pool M

I-6   Value k for nearest neighbor distance metric.

       Commonly, $k = \sqrt{N + \bar{N}}$

S-1   Create a random initial population pool

S-2   Set the initial archive as an empty set

S-3   Create the union set by combining together the states of the archive and population pool. The union set and archive set contain only unique members, whereas the population pool may contain duplicates.

S-4   For each state i in the union count the number of states dominated by i. Let p_i be the number of states dominated by state i.

       For each state i, compute

$$S_i = \sum_{i<j} p_j$$

S-5

       Given a state i, find all states j that dominate state i ($i < j$). Sum the value of p_j for each state j that dominates state i.

S-6   For each state i in the union count the distance to every other state in the union. Sort the distances to the other states by increasing order of distance. Let d_{ij} be the distance from state i to the jth state in rank order with smaller distances associated with smaller values of j.

       Compute the distance metric

S-7
$$D_i = \frac{1}{d_{ik} + 2}$$

       where k is set in I-6.

S-8
Compute the fitness for the ith state as

$$F_i = S_i + D_i$$

S-9
Determine all states in the union that are not dominated by any other state. These states be identified from the previous computations in S-5 as the states where $S_i = 0$.

S-10
Replace the archive set with the set of unique states in the union that are not dominated.

1. If the resulting number of states in the archive set is exactly $\bar{N}$, then continue to S-11.
2. If the resulting number of states in the archive set is smaller than $\bar{N}$, then add in the states from the union set of the previous generation until the current archive has $\bar{N}$ states.
3. If the resulting number of states in the archive set is larger than $\bar{N}$, then remove states from the archive by:
   a. Identify the state with the smallest value of d_{i1} (distance to its nearest neighbor). If multiple states are tied for the same smallest value, break the tie by considering the d_{i2}, etc.
   b. Remove from the archive the state identified in substep (3)(a) above.

S-11
Perform binary tournament selection with replacement on the new archive set to create a mating pool.

1. Clear the previous mating pool.
2. Choose two states from the archive set.
3. Copy the state with the *lower* fitness to the mating pool.
4. Do not remove these states from the archive set.
5. Repeat from substep (2) above until the population pool has M states.

S-12
Clear the population pool and apply recombination / mutation on the mating pool to create a new population pool.

S-13
Repeat from S-3 as many times as desired or until sufficient nondominated points are identified.

O-1
The list of states in the archive set is the approximation to the Pareto frontier.

Algorithm XXXVII: SPEA2 may be used to identify the Pareto frontier.

8.8 Nondominated Sorting Genetic Algorithm II

The nondominated sorting genetic algorithm II (NSGA-II) is an enhancement on the original NSGA to include elitism and improves the scaling behavior of the algorithm as the number of states increases. NSGA -II is an evolutionary algorithm for identifying the Pareto frontier in a multiobjective optimization problem.

The algorithm proceeds iteratively based on a parent and child states. At each generation (iteration), a set of child states is constructed by applying elitism on the parent states in conjunction with recombination and mutation. The child states are combined with the parent states, and a series of filtered sets are constructed. The filtered sets are used to construct the next set of parent states, which in turn create the next set of child states, etc.

The algorithm begins by constructing an initial parent population P_0 of N randomly selected states. During the setup phase, a set of child N states Q_0 is constructed using binary tournament selection with replacement, recombination, and mutation.

A comparison operator is required in order to apply elitism. NSGA -II uses a two-component comparison operator. The first component is a 'non-domination level' and the second component is a 'crowding distance'. For the non-domination level, lower scores are more desirable while higher scores are preferred for the crowding distance. For a given state, let ND_i be the non-domination and C_i be the crowding distance. The comparison operator between two states i and j is the total ordering operator $S \times S \to \{0,1\}$

$$S_i <_n S_j \to \left(ND_i < ND_j\right) \vee \left[\left(ND_i = ND_j\right) \wedge \left(C_i > C_j\right)\right] \qquad 8.73$$

In other words, we say that state i is less than state j if the non-domination level for state i is less than the non-domination level for j, or if the non-domination levels for the states are equal, and the crowding distance for state i is greater than the crowding level for j.

The operator $<_n$ is a total ordering operator. Given any two states S_i and S_j, we must have one of three results. Either $S_i <_n S_j$ is true, or $S_j <_n S_i$ is true, or $S_i = S_j$ (meaning that both $ND_i = ND_j$ and $C_i > C_j$). However, it may be true that the underlying states are different in that they have different component values, but both the non-dominance level and crowding distance are the same.

The comparison operator uses the values of the non-domination level and crowding distance to compare two states. The details for computing these metrics is provided below.

Non-Domination Level – The non-domination level is an integer value on the range $[1 \ldots N]$ where N is the total number of states under consideration. Let $T = P \cup C$ (the union of the current parent and child states) be a set of states with N members. The non-domination level for each state is computed in an iterative algorithm.

Each state is compared against every other state in T. Consider state $T_i \in T$. Count the number of states $T_j \in T$ where $T_i < T_j$ (dominance count). In addition, compile a list of states T_j where $T_j < T_i$ (dominated list). Keep in mind that dominance here is with respect to the usual Pareto comparison of the states in objective space not the comparison operator $<_n$. This step computes the value of F_i that is later used in the comparison operator $<_n$.

Once we have computed the dominance count and dominated list for each state in T, we identify all states where the dominance count is zero. These are all states that are not dominated by any other state. This is the approximation to the Pareto frontier from the states in T.

Next, remove each of the nondominated states from T. As each state is removed, examine the dominated list. For each state in the dominated list, reduce its dominance count by one. After all nondominated states are removed from T, the dominance count for the remaining states is correct with respect to the remaining states.

Finally, repeat the above process iteratively. At each iteration, a new set of nondominated states is identified and removed. The process is continued until all states are removed from T. As a state is removed, we track the iteration when the state was removed.

The above process results in a series of states. Let $\mathcal{F}_1$ be the set of states removed on the first iteration, $\mathcal{F}_2$ the states removed on the second iteration, etc. The non-domination level is the value of n where the state is assigned to the set $\mathcal{F}_n$.

The set $\mathcal{F}_1$ is the set of all non-dominated states that exist in T. In any set of states, there must be at least one non-dominated state. This may be proved through induction as shown below.

Proof:

Let $A^{(N)}$ be a poset of N states with a comparison operator $<$ and assume that there is at least one state $A_k \in A$ that is not dominated by any other state in $A^{(N)}$. Let $A^{(N+1)} = A^{(N)} \cup A_{N+1}$ be the set $A^{(N)}$ with an additional state A_{N+1}. If A_{N+1} is not dominated by any state in $A^{(N)}$, then $A^{(N+1)}$ has at least one non-dominated state. If it is dominated by at least one state in $A^{(N)}$, then $A^{(N+1)}$ cannot dominate A_k because the poset comparison operator is transitive. In this case, A_k is still not dominated by any state in $A^{(N+1)}$, and again $A^{(N+1)}$ must have at least one non-dominated state.

Thus, if $A^{(N)}$ has at least one non-dominated state, then $A^{(N+1)}$ must also have at least one non-dominated state. Finally, $A^{(1)}$ (a set with exactly one state) has a state that is not dominated by any other state. Thus, there exists a set $A^{(1)}$ that has at least one non-dominated states, and from the above we know that every set $A^{(N+1)}$ for all values of $N \geq 0$ must have at least one non-dominated state. ∎

If T is non-empty, there must be at least one non-dominated state in T. Thus, $|\mathcal{F}_1| > 0$. If the set $T_2 = T - \mathcal{F}_1$ is also not empty, then there must be at least one non-dominated state in T_2, so $|\mathcal{F}_2| > 0$. If the set $T_3 = T_2 - \mathcal{F}_2$ is also not empty, then there must be at least one non-dominated state in T_3, so $|\mathcal{F}_3| > 0$. The process may be continued until we reach a set T_{m+1} that is empty. If T_{m+1} is the first empty set in the sequence, then $|\mathcal{F}_k| > 0$ for $k = 1 \ldots m$. Since each T_k must have at least one state, the maximum value for m is N.

Each state in T is assigned to exactly one set $\mathcal{F}_k$. The value of k assigned to a state is the non-dominance level for the state.

Crowding Distance – The crowding distance examines states as grouped in one of the sets $\mathcal{F}_k$ identified during the computation of the non-dominance level. The crowding distance is computed for every state in T, but this is accomplished by examining each $\mathcal{F}_k$ individually.

For a particular set $\mathcal{F}_k$, we order the states by the value of each component. Let $\mathcal{F}_k^{(1)}$ be the sequence formed by ordering the states $\mathcal{F}_k$ by the value of the first component. Let $\mathcal{F}_k^{(1)(j)}$ be the jth state in the sequence, where the sequence is ordered such that $\mathcal{F}_k^{(1)(1)}$ has the smallest value for the first component, $\mathcal{F}_k^{(1)(2)}$ is the second smallest value, etc. In general, $\mathcal{F}_k^{(i)(j)}$ is jth state the sequence ordering of $\mathcal{F}_k$ by the ith component. Finally, let $\mathcal{F}_k^{(i)(j)}{}_p$ be the value of component p of the state.

Let $|\mathcal{F}_k| = M$ ($\mathcal{F}_k$ has M states), and let l be the number of components in the objective function. The crowding distance for each state $F \in \mathcal{F}_k$ is

$$ C = \sum_{n=1}^{l} \left(\mathcal{F}_k^{(n)(j+1)}{}_n - \mathcal{F}_k^{(n)(j-1)}{}_n \right) \qquad 8.74 $$

This is not valid when the value of a component is either the highest or lowest. In these cases, the crowding distance is set to ∞.

This expression looks more intimidating then it is in practice. Essentially, we take the set $\mathcal{F}_k$ and order it by one of it's components. The highest and lowest stated in the sorting are assigned a crowding distance of ∞. All other states have their crowding distance increased by the difference between the values of the component for the two states that bound the state. We repeat this for each component sorting of $\mathcal{F}_k$. Thus, the resulting crowding distance is large when a state is far from all other states, and small when the state is nearby many states.

Finally, we repeat this process for each of the $\mathcal{F}_k$'s identified in the non-dominance level computation above. After repeating for each $\mathcal{F}_k$, every state is assigned a positive value for the crowding distance.

At this point we have assigned each state a non-dominance level and a crowding distance. Moreover, the set of parents and children together are partitioned into a series of sets $\mathcal{F}_k$. From this, we are ready to create the next generation.

The next generation of parents is constructed from the sets $\mathcal{F}_k$. Let $P_{(n+1)}$ represent the next generation of parents and $C_{(n+1)}$ be the next generation of children. Thus, the sets $\mathcal{F}_k$ were constructed from $T = P_{(n)} \cup C_{(n)}$.

Initially set $P_{(n+1)} = \mathcal{F}_1$. If $|\mathcal{F}_1| \geq N$ we are done (N is the minimum number of parents in a generation). Otherwise, set $P_{(n+1)} = \mathcal{F}_1 \cup \mathcal{F}_2$. If $|\mathcal{F}_1 \cup \mathcal{F}_2| \geq N$, then the generation of parents is complete. Otherwise, continue to add in sets $\mathcal{F}_k$ until there are a sufficient number of states in the parents set.

Once $P_{(n+1)}$ is constructed, use the operator $<_n$ from 8.73 to order the states in $P_{(n+1)}$. Since $<_n$ is a total ordering, the states may be sorted into a unique order up to configuration differences in the case where two different states have the same non-dominance level and the same crowding distance. When two or more states have the same values for both of these metrics, then there is no unique ordering for this subset of states. However, in practice, since the crowding distance is a real-valued metric, it is rarely the case that two different states have the exact same value of the crowding distance.

Let $P_{(n+1)}{}^{(j)}$ be the sequence of states from $P_{(n+1)}$ sorted under $<_n$ with $j = 1$ assigned to the best state. Truncate this sequence to the first N states. Use selection, crossover, and mutation on the resulting sequence to create the next generation of child states $C_{(n+1)}$ where $|C_{(n+1)}| = N$.

NSGA-II

I-1   Independent variables $\vec{x}$

I-2   Objective function $\vec{F}(\vec{x})$ with components $f_n(\vec{x})$

I-3   Evolutionary method for generating a new population based on the fitness values for the states in the current population.

I-4   Size of the population pool N

S-1   Create a random initial population pool of N parent states $P_{(0)}$

S-2   Create an initial population pool of N child states $C_{(0)}$ from $P_{(0)}$ by using binary tournament selection with replacement, crossover, and mutation

S-3   Create the set $T = P_{(n)} \cup C_{(n)}$

S-4   For each state $T_k \in T$:

1. Count the number of states in T that dominate T_k (non-dominance level)
2. Create a list of all states T_s where T_k dominates T_s (dominance list)

S-5   For each state $T_k \in T$ where the non-dominance level is zero:

1. Remove the state T_k
2. Add T_k to the set $\mathcal{F}_1$
3. For each state T_l on the dominance list for T_k, subtract one from the non-dominance level of T_l.

Repeat these steps again, but in step (2) assign the states to the set $\mathcal{F}_2$. Continue repeating this process until T is empty. For each iteration, assign the states to a new set $\mathcal{F}_n$.

S-6   Set $P_{n+1} = \mathcal{F}_1$. If $|P_{n+1}| \geq N$ continue to S-7.

Otherwise set $P_{n+1} = \mathcal{F}_1 \cup \mathcal{F}_2$. If $|P_{n+1}| \geq N$ continue to S-7.

Otherwise, continue to add additional sets $\mathcal{F}_k$ until $|P_{n+1}| \geq N$.

S-7   Let k_{max} be the maximum value for $\mathcal{F}_k$ required in S-6. For each $\mathcal{F}_k$ where $k = 1 \ldots k_{max}$ compute the crowding distance:

1. Initialize the crowding distance for each state in $\mathcal{F}_k$ to zero
2. For each component i of the objective function
   a. Sort the states in $\mathcal{F}_k$ by the value of component i
   b. Let $\mathcal{F}_k^{(i)}$ represent the sequence formed from sorting $\mathcal{F}_k$ by the ith component

    c.   Let $\mathcal{F}_k^{(i)(j)}$ be the jth state in the ordered sequence where $j = 1$ is the lowest value of the component and the sequence increases in value

    d.   Let $\mathcal{F}_k^{(i)(j)}{}_n$ be the value of the nth component of the of the jth state in the sorted sequence $\mathcal{F}_k^{(i)}$

    e.   For the first and last state in the sequence $\mathcal{F}_k^{(i)}$, set the crowding distance to infinity

    f.   For every other state in the sequence $\mathcal{F}_k^{(i)}$, increment the crowding distance for the state by the value

$$\mathcal{F}_k^{(i)(j+1)}{}_i - \mathcal{F}_k^{(i)(j-1)}{}_i$$

    g.   Repeat these steps for each component of $\mathcal{F}_k$

  3. Repeat these steps for each $\mathcal{F}_k$

S-8

Sort P_{n+1} using the total ordering operator

$$S_i <_n S_j \rightarrow \left(ND_i < ND_j\right) \vee \left[\left(ND_i = ND_j\right) \wedge \left(C_i > C_j\right)\right]$$

where $S_i, S_j \in P_{n+1}$, ND_i is the non-dominance level for S_i as computed in S-5, and C_i is the crowding distance as computed in S-7. Let $P_{n+1}^{(i)}$ be the sorted sequence where $i = 1$ corresponds to the best values (ND_i is lowest and C_i is highest).

S-9

Truncate $P_{n+1}^{(i)}$ to the first N states in the sequence

S-10

Perform binary tournament selection with replacement on the new archive set to create the next generation child set $C_{(n+1)}$.

  1. Choose one or more states from $P_{n+1}^{(i)}$

  2. Use selection, recombination, mutation to generate a new state based on the states selected

  3. Do not remove these states from the archive set

  4. Add the new state to $C_{(n+1)}$

  5. Continue from (1) until $\left|C_{(n+1)}\right| = N$

S-11

Repeat from S-3 as many times as desired or until sufficient nondominated points are identified.

O-1

The list of nondominated states in $T = P_{(n+1)} \cup C_{(n+1)}$ in the archive set is the approximation to the Pareto frontier.

Algorithm XXXVIII: NSGA-II may be used to identify the Pareto frontier.

8.9 Directed Search Domain

The directed search domain (DSD) is a technique for identifying the Pareto frontier for a multiobjective optimization problem similar to the methods of normal-boundary intersection and normal constraint. The directed search domain identifies the same utopia hyperplane, but looks for solutions within a cone whose vertex is the normal to the hyperplane.

Let the optimization function have the form

$$\vec{F}(\vec{x}) = \begin{bmatrix} f_1(\vec{x}) \\ f_2(\vec{x}) \\ \vdots \\ f_n(\vec{x}) \end{bmatrix}$$

8.75

DSD begins by using single objective optimization to find the maximum value of each component of the objective function independently. Let $\vec{x}_i^*$ be the optimum parameter vector for the component $f_i(\vec{x})$, and set $\vec{F}(\vec{x}_i^*) = \vec{F}_i^*$. The points $\vec{F}(\vec{x}_i^*)$ are called the anchor points.

The points are used to transform the objective space using the Miettinen transform. This transform scales each component of the objective function independently so that the transformed components are on the range $[0,1]$ for the n single objective optima.

The Miettinen transform is based on the Nadir point:

$$\vec{N} = \langle N_1, N_2, \dots, N_n \rangle$$

8.76

where

$$N_i = min[f_i(\vec{x}_1^*), f_i(\vec{x}_2^*), \dots, f_i(\vec{x}_n^*)]$$

8.77

Once the Nadir point is identified, the components of the objective function are transformed as

$$\bar{f}_i(\vec{x}) = \frac{f_i(\vec{x}) - N_i}{f_i(\vec{x}_i^*) - N_i}$$

8.78

Under the Mietten transformation, evaluating each component over the set of optimum parameter vectors $\vec{x}_n^*$ results in values on the range $[0,1]$. Each component is transformed independently, so the transformation merely scales each component.

The transformed objective function is

$$\vec{F}(\vec{x}) = \begin{bmatrix} \vec{f}_1(\vec{x}) \\ \vec{f}_2(\vec{x}) \\ \vdots \\ \vec{f}_n(\vec{x}) \end{bmatrix}$$

8.79

The DSD method uses the utopia hyperplane similar to the normal-boundary intersection and normal constraint methods. The utopia hyperplane is the $n-1$ dimensional subspace containing the anchor points.

Under the Mietten transformation, the utopia vectors $\vec{U}^i$ are each unit vectors where all components are zero except one. The utopia point in the transformed coordinates is the vector $\langle 1, 1, \ldots, 1 \rangle$.

From the transformed utopia vectors, we construct points on the utopia hyperplane such that

$$\vec{X}^j = \sum_{i=1}^{n} \alpha_{ij} \hat{\varepsilon}_i$$

8.80

where

$$\sum_{i=1}^{n} \alpha_{ij} = 1$$

8.81

The values α create a set of vectors $\vec{\alpha}^j$ that point to points on the utopia hyperplane.

The normal to the utopia hyperplane is the vector

$$\hat{n} = \frac{1}{\sqrt{n}} \langle 1, 1, \ldots, 1 \rangle$$

8.82

The DSD method maximizes the sum of the components of the objective function

$$max \sum_{i=1}^{n} \vec{F}_i(\vec{x})$$

8.83

subject to the constraint that the point $\vec{F}(\vec{x})$ must lie within a cone of angle γ with vertex on the point $\vec{X}^j$ and axis parallel to the normal to the hyperplane. This constraint may be enforced from the dot product:

$$\left[\vec{F}(\vec{x}) - \vec{X}^j \right] \cdot \hat{n}$$

8.84

The dot product may be evaluated in two ways. First, compute the dot product by multiplying the components of the vectors together and summing:

$$\left[\vec{F}(\vec{x}) - \vec{X}^j\right] \cdot \hat{n} = \frac{1}{\sqrt{n}} \sum_{i=1}^{n} \left(F_i(\vec{x}) - \alpha_{ji}\right) \qquad 8.85$$

This may also be written in terms of the angle θ between the vectors:

$$\left[\vec{F}(\vec{x}) - \vec{X}^j\right] \cdot \hat{n} = \left|\vec{F}(\vec{x}) - \vec{X}^j\right| |\hat{n}| \cos\theta \qquad 8.86$$

$$= \sqrt{\sum_{i=1}^{n} \left(F_i(\vec{x}) - \alpha_{ji}\right)^2} \cos\theta \qquad 8.87$$

Equating the two methods,

$$\frac{1}{\sqrt{n}} \sum_{i=1}^{n} \left(F_i(\vec{x}) - \alpha_{ji}\right) = \sqrt{\sum_{i=1}^{n} \left(F_i(\vec{x}) - \alpha_{ji}\right)^2} \cos\theta \qquad 8.88$$

or,

$$\cos\theta = \frac{\sum_{i=1}^{n} \left(F_i(\vec{x}) - \alpha_{ji}\right)}{\sqrt{n \sum_{i=1}^{n} \left(F_i(\vec{x}) - \alpha_{ji}\right)^2}} \qquad 8.89$$

Thus, if we are constrained to a cone of angle γ, then

$$\cos\theta \geq \cos\gamma \qquad 8.90$$

With this the constraint becomes

$$\frac{\sum_{i=1}^{n} \left(F_i(\vec{x}) - \alpha_{ji}\right)}{\sqrt{n \sum_{i=1}^{n} \left(F_i(\vec{x}) - \alpha_{ji}\right)^2}} \geq \cos\gamma \qquad 8.91$$

For problems with three of more components, there may be points on the Pareto surface that do not have a normal projection onto the utopia hyperplane. These points are never identified under the traditional NIB and NC methods.

However, DSD may identify these points by varying the angle of the constraint cone so that the axis of the cone is no longer parallel to the normal to the hyperplane. If the cone has axis parallel to the vector $\hat{c}$, the dot product is

$$\left[\vec{F}(\vec{x}) - \vec{X}^j\right] \cdot \hat{c} = \sum_{i=1}^{n} \left(F_i(\vec{x}) - \alpha_{ji}\right)c_i \qquad 8.92$$

so the constraint becomes

$$\frac{\sum_{i=1}^{n} \left(F_i(\vec{x}) - \alpha_{ji}\right)c_i}{\sqrt{\sum_{i=1}^{n} \left(F_i(\vec{x}) - \alpha_{ji}\right)^2}} \geq \cos\gamma \qquad 8.93$$

Directed Search Domain

I-1   Independent variables $\vec{x}$

I-2   Objective function $\vec{F}(\vec{x})$ with components $f_n(\vec{x})$

I-3   Evolutionary method for generating a new population based on the fitness values for the states in the current population.

I-4   Number of hyperplane points to consider N

I-5   Method for selecting a random angle γ on the range $\left[0, \frac{\pi}{4}\right)$

I-6   Method for selecting a random directional vector $\hat{c}$

I-7   Method for optimizing a single unconstrained objective function

S-1   Compute the global optimum for each individual component of the objective function. Let $\vec{x}_i^*$ designate the global optimum for $f_i(\vec{x})$

S-2   Compute the Nadir point $\vec{N} = \langle N_1, N_2, \ldots, N_n \rangle$ where
$$N_i = min[f_i(\vec{x}_1^*), f_i(\vec{x}_2^*), \ldots, f_i(\vec{x}_n^*)]$$

S-3   Transform the coordinates using
$$\bar{f}_i(\vec{x}) = \frac{f_i(\vec{x}) - N_i}{f_i(\vec{x}_i^*) - N_i}$$

S-4   Create a lattice of equally spaced points on the utopia hyperplane
$$\vec{\bar{X}}^j = \sum_{k=1}^{n} \alpha_{kj} \hat{e}^k \qquad \sum_{k=1}^{n} \alpha_{kj} = 1 \qquad |\alpha_{kj}| \leq 1$$

S-5   For every $\vec{\bar{X}}^j$, optimize the single objective function
$$max \sum_{i=1}^{n} \vec{F}_i(\vec{x})$$
subject to the constraint
$$\frac{\sum_{i=1}^{n}(F_i(\vec{x}) - \alpha_{ji})c_i}{\sqrt{\sum_{i=1}^{n}(F_i(\vec{x}) - \alpha_{ji})^2}} \geq cos\,\gamma$$
where γ is a random angle on the range $\left[0, \frac{\pi}{4}\right)$ and $\hat{c}$ is a directional vector in the direction of the axis of the constraint cone.

O-1   The list of nondominated states in $T = P_{(n+1)} \cup C_{(n+1)}$ in the archive set is the approximation to the Pareto frontier.

Algorithm XXXIX: The directed search domain method may be used to identify the frontier.

8.10　　Random Weight Genetic Algorithm

The random weight genetic algorithm (RWGA) combines the weighted sum method with a genetic algorithm. At each iteration, a new weight vector is created, and each iteration applies the states to the weight vector. This allows the RWGA method to search many weights in a single run of the algorithm.

Let the optimization function have the form

$$\vec{F}(\vec{x}) = \begin{bmatrix} f_1(\vec{x}) \\ f_2(\vec{x}) \\ \vdots \\ f_n(\vec{x}) \end{bmatrix} \qquad \qquad 8.94$$

To begin, the RWGA method creates an initial population of N states chosen at random. These initial states are placed into an archive. As states are added to the archive, the added state is checked against the states in the archive. If the state is not dominated by any state in the archive, the state is copied to the archive. If the state is dominated by any state in the archive, the state is rejected. If the state is accepted, any state in the archive that is dominated by the newly added state is removed from the archive. Thus, the archive contains only nondominated states and is the approximation to the Pareto frontier.

For each iteration of the algorithm, a weight vector is chosen at random where

$$\sum_{i=1}^{n} \alpha_i = 1 \quad |\alpha_i| = 1 \qquad \qquad 8.95$$

Based on this random vector, a fitness is computed for each state. Let $\vec{F}(\vec{x}_j)$ be the value of the objective function for state j. The fitness for the state is

$$f_j = \sum_{i=1}^{n} \alpha_i F_i(\vec{x}_j) \qquad \qquad 8.96$$

Let f_{min} be the minimum value of the fitness from all the states. Assign eash state a selection probability

$$p_j = \frac{f_j - f_{min}}{\sum_{k=0}^{N}[f_k - f_{min}]} \qquad \qquad 8.97$$

Finally, select states from the population based on the selection probability and apply recombination and mutation to create a set of child states. Add these states to the archive as above. Randomly remove n_E states from the child population, and add n_E states from the archive. Then repeat for the next iteration.

Random Weighted Genetic Algorithm

I-1   Independent variables $\vec{x}$

Objective function

I-2

$$\vec{F}(\vec{x}) = \begin{bmatrix} f_1(\vec{x}) \\ f_2(\vec{x}) \\ \vdots \\ f_n(\vec{x}) \end{bmatrix}$$

Method for choosing a random weight vector $\vec{\alpha}$ where

I-3

$$\sum_{i=1}^{n} \alpha_i = 1 \quad |\alpha_i| \leq 1$$

I-4   Population size N

I-5   Population to remove and replace from the archive n_e

S-1   Create an initial population of states chosen at random

S-2   Generate a random weight vector $\vec{\alpha}$

Compute the fitness value for each state as

S-3

$$f = \sum_{i=1}^{n} \alpha_i F_i(\vec{x})$$

S-4   Compute the minimum fitness across all states f_{min}

Compute the selection probability for each state as

S-5

$$p_i = \frac{f_i - f_{min}}{\sum_{k=0}^{N}[f_k - f_{min}]}$$

S-6   Select states according to the selection probability, apply crossover and mutation to create N child states

S-7   Add the child states to the archive and retain all nondominated solutions in the archive

S-8   Remove n_E states at random from the child population. Replace these with randomly selected states from the archive

O-1   The archive represents the approximation to the Pareto frontier

Algorithm XL: The random weighted sum genetic algorithm method may be used to identify the Pareto frontier.

8.11 Utopia Lattice Weighted Sum

The utopia lattice weighted sum (ULWS) combines the weighted sum method with a method for generating weights. In ULWS, the weights are generated by identifying the utopia hyperplane, then constructing a uniform lattice of points on the utopia hyperplane. The position of the lattice points corresponds to the weights used in the weighted sum method.

Let the optimization function have the form

$$\vec{F}(\vec{x}) = \begin{bmatrix} f_1(\vec{x}) \\ f_2(\vec{x}) \\ \vdots \\ f_n(\vec{x}) \end{bmatrix} \qquad 8.98$$

ULWS begins by computing the optimum value for each component individually. Let $\vec{x}_i^*$ be the optimum parameter vector for the component $f_i(\vec{x})$, and set $\vec{F}(\vec{x}_i^*) = \vec{F}_i^*$.

From these values, the Nadir point may be identified as

$$\vec{N} = \langle N_1, N_2, \ldots, N_n \rangle \qquad 8.99$$

where

$$N_i = min[f_i(\vec{x}_1^*), f_i(\vec{x}_2^*), \ldots, f_i(\vec{x}_n^*)] \qquad 8.100$$

The Nadir point is then used to apply the Miettinen transform to each of the components of the objective function. Each component is transformed as

$$\bar{f}_i(\vec{x}) = \frac{f_i(\vec{x}) - N_i}{f_i(\vec{x}_i^*) - N_i} \qquad 8.101$$

Under this transformation, the vectors to each of the component minimums is just unit vector $\hat{\varepsilon}_i$ for the component. Based on this, the utopia hyperplane may be defined as span of the $n - 1$ vectors

$$\vec{H}^i = \hat{\varepsilon}_i - \hat{\varepsilon}_n \qquad 8.102$$

Equally distributed points in the hyperplane may be found as

$$\vec{X}^j = \sum_{k=1}^{n} \alpha_{kj} \vec{H}^i \qquad \sum_{k=1}^{n} \alpha_{kj} = 1 \quad |\alpha_{kj}| \le 1 \qquad 8.103$$

The values of α_{kj} may be chosen to create a uniform distribution of points in the utopia hyperplane. The values of α_{kj} associated with a particular vector $\vec{X}^j$ are used as the weight vector in the weighted sum method.

Utopia Lattice Weighted Sum

I-1   Independent variables $\vec{x}$

I-2   Objective function

$$F(\vec{x}) = \begin{bmatrix} f_1(\vec{x}) \\ f_2(\vec{x}) \\ \vdots \\ f_n(\vec{x}) \end{bmatrix}$$

I-5   Population size N

S-1   Compute the global optimum for each individual component of the objective function. Let $\vec{x}_i^*$ designate the global optimum for $f_i(\vec{x})$

S-2   Compute the Nadir point $\vec{N} = \langle N_1, N_2, \dots, N_n \rangle$ where
$$N_i = min[f_i(\vec{x}_1^*), f_i(\vec{x}_2^*), \dots, f_i(\vec{x}_n^*)]$$

S-3   Transform the coordinates using
$$\bar{f}_i(\vec{x}) = \frac{f_i(\vec{x}) - N_i}{f_i(\vec{x}_i^*) - N_i}$$

S-4   Create a lattice of equally spaced points on the utopia hyperplane
$$\vec{X}^j = \sum_{k=1}^{n} \alpha_{kj}(\hat{\varepsilon}_i - \hat{\varepsilon}_n) \quad \sum_{k=1}^{n} \alpha_{kj} = 1 \quad |\alpha_{kj}| \leq 1$$

S-5   For each value j, maximize the single objective function
$$g(\vec{x}) = \sum_{i=1}^{n} \alpha_{ij} f_i(\vec{x})$$

Let $\vec{x}_j^*$ designate this maximum.

S-6   Compute the value of $\vec{F}(\vec{x}_j^*)$

S-7   Add $\vec{F}(\vec{x}_j^*)$ to the archive and retain all nondominated solutions in the archive

O-1   The archive represents the approximation to the Pareto frontier

Algorithm XLI: The utopia lattice weighted sum method may be used to identify the Pareto frontier.

8.12   Vector Evaluated Genetic Algorithm

The vector evaluated genetic algorithm (VEGA) is the first genetic algorithm designed to identify the Pareto frontier for a multiobjective optimization problem. VEGA optimizes each component of the objective function and uses a genetic algorithm to perform selection, recombination, and mutation to evolve a set of states toward the Pareto frontier.

Let the optimization function have the form

$$\vec{F}(\vec{x}) = \begin{bmatrix} f_1(\vec{x}) \\ f_2(\vec{x}) \\ \vdots \\ f_n(\vec{x}) \end{bmatrix} \qquad\qquad 8.104$$

Initially, VEGA starts with a population of N states that are chosen at random. During each iteration, the population is randomly sorted into a sequence. The sequence is divided into n groups (n is the number of components in the objective function).

The fitness for the states in the kth group is the value of the kth component of the state. The first group of states has a fitness equal to the value of the first component of the objective, the second group of states has fitness equal to the second component of the objective, etc.

This way each fitness component appears somewhere. Also, since the states are randomly sequenced every generation, each state has an opportunity to input each objective component at some point during the iterative process.

Figure 134 shows the VEGA method applied to the multiobjective optimization problem from equation 8.32. Similarly, Figure 135 shows application of the VEGA method to the problem from equations 8.71-8.72.

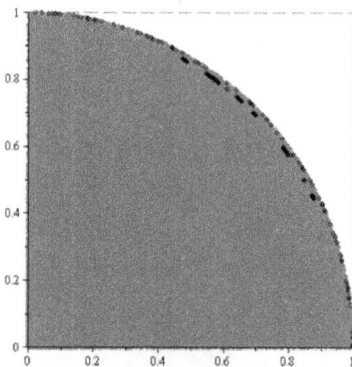

Figure 134: VEGA method applied to the quarter circle problem from 8.32.

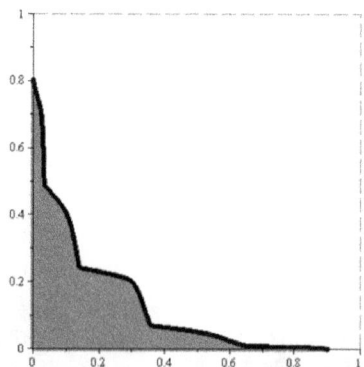

Figure 135: VEGA method applied to the problem from 8.71-8.72.

Once the fitness values are assigned, selection, recombination, and crossover may be used to create the next generation population. The new states are added to an archive set that maintains a list of the nondominated states found thus far. The states in the archive are adjusted according to the states in the new population to keep the archive states nondominated. The archive is the approximation to the Pareto frontier.

Vector Evaluated Genetic Algorithm

I-1   Independent variables $\vec{x}$

Objective function

I-2
$$\vec{F}(\vec{x}) = \begin{bmatrix} f_1(\vec{x}) \\ f_2(\vec{x}) \\ \vdots \\ f_n(\vec{x}) \end{bmatrix}$$

I-3   Population size N, where $N \bmod n \equiv 0$

S-1   Create an initial population of N states chosen at random

S-2   Create a sequence from the population where the sequence order is random

S-3   Divide the population into n equal groups

S-4   Set the fitness for the states in the kth group equal to the value of $F_k(\vec{x})$ for the state

S-5   Compute the minimum fitness across all states f_{min}

Compute the selection probability for each state as

S-6
$$p_i = \frac{f_i - f_{min}}{\sum_{k=0}^{N}[f_k - f_{min}]}$$

S-7   Use selection, recombination, and mutation to create a new population

S-8   Add the new states to an archive maintaining only nondominated states

O-1   The archive represents the approximation to the Pareto frontier

Algorithm XLII: The vector evaluated genetic algorithm method may be used to identify the Pareto frontier.

8.13   Pareto Envelope Based Selection Algorithm

The Pareto envelope based selection algorithm (PESA-II) is a genetic algorithm method for identifying the Pareto frontier. However, unlike many other genetic algorithm methods, the fitness selection process is not based on the individual states. Rather, PESA-II divides the operational space into a series of regions, and each region is assigned a fitness. Regions are selected rather than individuals. Once a region is selected, one of the states in the region is chosen and recombination and mutation is applied to the selected state(s).

Let the optimization function have the form

$$\vec{F}(\vec{x}) = \begin{bmatrix} f_1(\vec{x}) \\ f_2(\vec{x}) \\ \vdots \\ f_n(\vec{x}) \end{bmatrix} \qquad 8.105$$

PESA-II begins by creating an initial population of N states. These states are chosen at random.

The operational space is divided into a set of hypercubes by segmenting each component of the objective function into s regions. This divides the operational space into n^s hypercubes where n is the number of components in the objective function.

The iterative process begins by counting the number of states in each cell. The number of states in a cell is assigned as the cell fitness. Cells that have no states are discarded. Based on this, cells are selected with lower fitness preferred.

The next generation of states is compiled by selecting cells based on their fitness and selecting a state randomly from the cell. Selected states undergo recombination and mutation to create new states for the next generation.

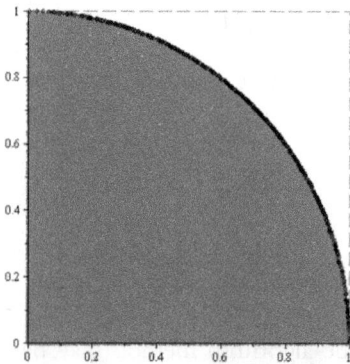

Figure 136: PESA-II method applied to the quarter circle problem from 8.32.

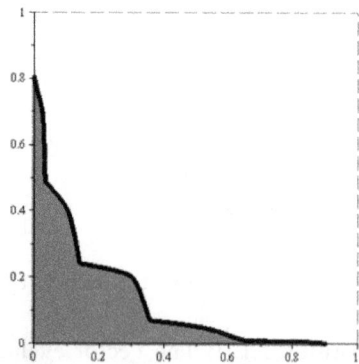

Figure 137: PESA-II method applied to the problem from 8.71-8.72.

Pareto Envelope Based Selection Algorithm II

I-1	Independent variables $\vec{x}$

I-2 Objective function

$$\vec{F}(\vec{x}) = \begin{bmatrix} f_1(\vec{x}) \\ f_2(\vec{x}) \\ \vdots \\ f_n(\vec{x}) \end{bmatrix}$$

I-3	Population size N
I-4	Number of segments s for dividing each component
S-1	Create an initial population of N states chosen at random
S-2	Divide the operational space into n^s cells, where n is the number of components for the objective function
S-3	Count the number of states in each cell
S-4	Set the fitness of a cell equal to the number of states in the cell
S-5	Discard all cells that have no states
S-6	Use binary tournament selection to select cells in proportion to their fitness
S-7	From a selected cell, randomly choose a state from the cell
S-8	Use recombination and mutation on the chosen state(s) to create a new state
S-9	Repeat from S-6 to create a new population of N states
S-10	Add the new states to the archive of nondominated states. Update the archive so that the set of states in the archive are all nondominated.
S-11	Set the population to the new population created from cell selection above
S-12	Repeat from S-3 for each iteration
O-1	The archive represents the approximation to the Pareto frontier

Algorithm XLIII: PESA-II is an evolutionary based algorithm that may be used to identify the Pareto frontier.

8.14 Dynamic Multiobjective Evolutionary Algorithm

The dynamic multiobjective evolutionary algorithm (DMEOA) is a multiobjective optimization method that combines principles of genetic algorithms and simulated annealing. The fitness score for the genetic algorithm is based in part on a component similar to that used in simulated annealing.

Let the optimization function have the form

$$\vec{F}(\vec{x}) = \begin{bmatrix} f_1(\vec{x}) \\ f_2(\vec{x}) \\ \vdots \\ f_n(\vec{x}) \end{bmatrix} \qquad 8.106$$

The algorithm begins by creating a population of N states at random. A three component fitness score is computed for each state. The fitness score combines a nondominance score, an 'entropy' dependent score, and a crowding metric. Each of these metrics is described below.

Nondominance Rank – For a state S, the nondominance rank is the count of the number of states that dominate S.

$$R_i = |\Omega_i| \qquad \Omega_i = \{X_j \mid X_j > X_i, 1 \le j \le n\} \qquad 8.107$$

where $>$ is the 'greater than' operator in the Pareto sense.

Entropy – The entropy for a state S is computed as

$$S_i = -p_i \log p_i \qquad p_i = \frac{e^{-R_i/T}}{\sum_{k=1}^{n} e^{-R_k/T}} \qquad 8.108$$

Crowding Distance - The crowding distance is an estimate of the smallest hypercube surrounding a state. The states are ordered by each component of the objective, and the distance between the two neighboring states is computed. These distances are summed together to compute the crowding distance.

Let $\mathcal{T}$ be the population of states under consideration. For each component of the objective function, the states are ordered by the value of the component. Let $\mathcal{T}^{(i)}$ be the sequence formed by ordering the states by the value of the ith component. Furthermore, let $\mathcal{T}^{(i)(j)}$ be the jth state in the sequence, where the sequence is ordered from low to high. Finally, let $\mathcal{T}^{(i)(j)}{}_p$ be the value of component p of the state. If n if the number of components in the objective function, the crowding distance is

$$C_i = \sum_{k=1}^{n} \left(\mathcal{T}^{(k)(j+1)}{}_k - \mathcal{T}^{(k)(j-1)}{}_k \right) \qquad 8.109$$

This is not valid when the value of a component is either the highest or lowest. In these cases, the crowding distance is set to ∞.

Putting these metrics together, the fitness for a state is computed as

$$F_i = R_i - TS_i - C_i \qquad \text{8.110}$$

Moreover, a temperature schedule is provided so that the temperature T of the system varies with the iteration. The temperature decreases with the iterations.

Once a fitness score is assigned to each state, a new population is created using selection, recombination, and mutation. The selection process for DMOEA is modified from the traditional approach used with genetic algorithms. The modified selection process is described below.

Selection Process – The selection process begins by sorting the list of states according to their fitness. States with high values of fitness are the worst states while states with low fitness are the best.

A child population of N individuals is created from the current population using the standard multi-parent crossover process of genetic algorithms. Fitness scores are computed for the daughters produced from multi-parent crossover. When computing the nondominance rank for the daughters, the daughter states are only compared with the original population, not the other daughter states.

The list of daughter states is compared with the sorted list of parents. Each daughter is compared with the current worst state in the population list. If a daughter is selected (the selection process is described below), the worst state on the list is eliminated, and the daughter is inserted.

The decision to select or discard a daughter is made by comparing the daughter with the current worst state in the population. The daughter state is selected if one of the three criteria below is met:

1. If the nondominance rank of the daughter is less than the nondominance rank of the worst parent
2. If the nondominance rank of the daughter is equal to the nondominance rank of the worst parent, and the crowding distance of the daughter is greater than the crowding distance of the parent.
3. Generate a uniform random number r on the range $[0,1)$. The daughter is selected if

$$exp\left[(R_{worst} - R_{daughter})/T\right] > r \qquad \text{8.111}$$

If none of these are met, the daughter is discarded.

Dynamic Multiobjective Evolutionary Algorithm

I-1   Independent variables $\vec{x}$

Objective function

I-2

$$\vec{F}(\vec{x}) = \begin{bmatrix} f_1(\vec{x}) \\ f_2(\vec{x}) \\ \vdots \\ f_n(\vec{x}) \end{bmatrix}$$

I-3   Population size N

S-1   Create an initial population of N states chosen at random. Designate the population of states as $\mathcal{T}$.

S-2   Assign each state a nondominance rank equal to the number of states that dominate the state. Set R_i as the nondominance rank for the ith state.

Compute the crowding distance C_i for each state:

1. Initialize the crowding distance for each state to zero
2. For each component i of the objective function
   a. Sort the states in by the value of component i
   b. Let $\mathcal{T}^{(i)}$ represent the sequence formed from sorting $\mathcal{T}$ by the ith component
   c. Let $\mathcal{T}^{(i)(j)}$ be the jth state in the ordered sequence where $j = 1$ is the lowest value of the component and the sequence increases in value

S-3   d. Let $\mathcal{T}^{(i)(j)}{}_n$ be the value of the nth component of the of the jth state in the sorted sequence $\mathcal{T}^{(i)}$
   e. For the first and last state in the sequence $\mathcal{T}^{(i)}$, set the crowding distance to infinity
   f. For every other state in the sequence $\mathcal{T}^{(i)}$, increment the crowding distance for the state by the value

   $$\mathcal{T}^{(i)(j+1)}{}_i - \mathcal{T}^{(i)(j-1)}{}_i$$

   g. Repeat these steps for each component of $\mathcal{T}$
3. Repeat these steps for each $\mathcal{T}$

Compute the entropy for each state as

S-4 $$S_i = -p_i \log p_i$$

where

$$p_i = \frac{e^{-R_i/T}}{\sum_{k=1}^{n} e^{-R_k/T}}$$

S-5 Compute the fitness for each state as

$$F_i = R_i - TS_i - C_i$$

S-6 Use binary tournament selection to create a child population based on the fitness of the population $\mathcal{T}$

S-7 Compute the metrics from S-2, S-3, and S-4 for each of the child states.

S-8 Compute the fitness for each of the child states.

S-9 Each of the child states is selected or rejected by comparing the metrics of the child state to the metrics of the current worst fitness of the parent population. A child is selected if any of the following conditions are met:

1. The nondominance rank of the child is less than the nondominance rank of the worst parent
2. If the nondominance rank of the child is equal to the nondominance rank of the worst parent, and the crowding distance of the child is greater than the crowding distance of the parent
3. If

$$exp\big[\big(R_{worst} - R_{daughter}\big)/T\big] > r$$

where r is a random number generated uniformly on the range $[0,1)$

If a child is selected, the worst parent is removed from the population and the child is added to the population and the new worst state is identified. The process is repeated for each of the child states.

S-10 Add the child population to the archive of nondominated states. Update the archive so that the set of states in the archive are all nondominated.

S-11 Set the population to the child population created

O-1 The archive represents the approximation to the Pareto frontier

Algorithm XLIV: DMOEA is an evolutionary based algorithm that may be used to identify the Pareto frontier.

8.15   Pareto Archived Evolution Strategy

The Pareto archived evolution strategy (PAES) is a genetic algorithm method for identifying the Pareto frontier. During each iteration, each state is mutated and the resulting state compared with the original state and an archive of nondominated states to determine if the new state should replace the original state or if the new state should be rejected.

If the new state dominates the original state, the new state replaces the original state. If the original state dominates the new state, the new state is rejected. If the states are Pareto incomparable, then the new state is compared with the archive of nondominated states. If the new state is not dominated by any state in the archive, then the new state is accepted. Otherwise the new state is rejected.

An archive of nondominated states is maintained during the execution of the algorithm. This archive is used both as the approximation to the Pareto frontier, and to determine if a new state should be selected when the new state and original state are Pareto incomparable.

An interesting aspect of this method is that since the method only used the mutation operator of the genetic algorithm, the entire method may be implemented using only a single state. The Pareto frontier may be discovered from this single state as the mutation operator moves the state through the operational space and the archive is used to place selection pressure on the state to continue to identify dominating states.

PAES is an attractive method due to the simplicity of implementation. No fitness computation is required, and the selection process only requires a Pareto comparison.

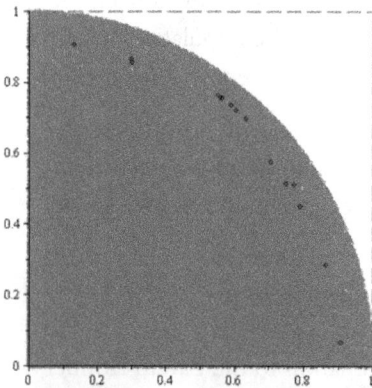

Figure 138: PAES method applied to the quarter circle problem from 8.32.

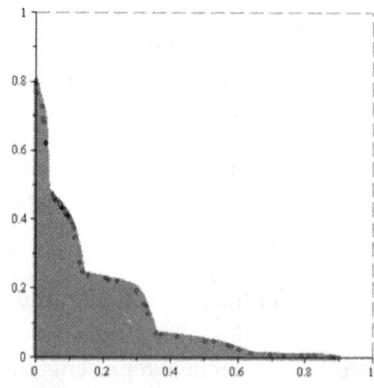

Figure 139: PAES method applied to the problem from 8.71-8.72.

Figure 138 shows the application of PAES to the problem from 8.32, while Figure 139 shows the application of PAES to the problem defined in equations 8.71-8.72. In both cases, PAES is able to find several points near the Pareto frontier for each problem.

However, in each case, the number of function evaluations is comparable with the previous methods. PAES does not seem to identify as many points as some of these other methods. However, the identified points are still fairly well distributed across the frontier.

Pareto Archive Evolution Strategy

I-1   Independent variables $\vec{x}$

        Objective function

I-2
$$\vec{F}(\vec{x}) = \begin{bmatrix} f_1(\vec{x}) \\ f_2(\vec{x}) \\ \vdots \\ f_n(\vec{x}) \end{bmatrix}$$

I-3   Population size N

S-1   Create an initial population of N states chosen at random

S-2   For each state, apply the mutation operator to create a new state.

S-3   If the new state is dominated by the original state, reject the new state and continue from S-2 for the next state

S-4   If the new state dominates the original state, then replace the original state with the new state, add the new state to the archive, and continue from S-2 for the next state

S-5   If neither state dominates the other, then compare the new state with the archive. If the new state is not dominated by any state in the archive, then replace the original state with the new state. If the new state is dominated by any state in the archive, reject the new state.

S-6   Continue from S-2 for the next state.

S-7   Repeat the entire process from S-2 for each iteration of the algorithm

O-1   The archive represents the approximation to the Pareto frontier

Algorithm XLV: PAES is an evolutionary based algorithm that may be used to identify the Pareto frontier.

8.16　　Niched Pareto Genetic Algorithm

The niched Pareto genetic algorithm (NPGA) is a genetic algorithm method for identifying the Pareto frontier that uses niching to apply pressure on the nondominated states to spread them out over the frontier.

Let the optimization function have the form

$$\vec{F}(\vec{x}) = \begin{bmatrix} f_1(\vec{x}) \\ f_2(\vec{x}) \\ \vdots \\ f_n(\vec{x}) \end{bmatrix} \qquad 8.112$$

NPGA begins by creating a population of initial states chosen at random over the operational space. For each state i in the population, find all other states j where the distance between the states in operational space is less than a distance parameter σ_{min}.

Let $\mathcal{D}_i$ be the set of all states that are within a distance of σ_{min} from state i, and let d_{ij} be the distance between states i and j. The fitness for state i is given by

$$f_i = \sum_{j \in \mathcal{D}_i} \left(1 - \frac{d_{ij}}{\sigma_{min}} \right) \qquad 8.113$$

The fitness is small for states that are isolated and large for states that have many near neighbors. Using this fitness in the selection process for the genetic algorithm puts selection pressure on the states to distribute themselves more evenly over the Pareto frontier.

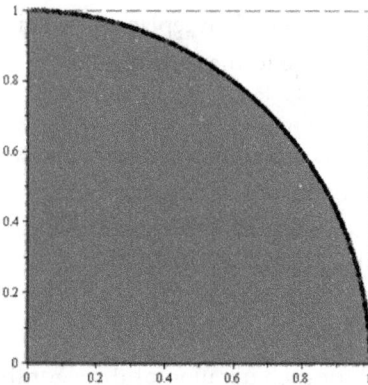

Figure 140: NPGA method applied to the quarter circle problem from 8.32.

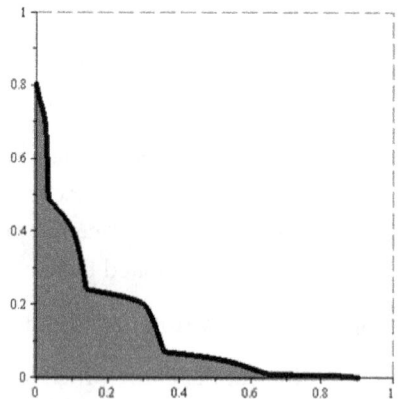

Figure 141: NPGA method applied to the problem from 8.71-8.72.

Figure 140 shows the application of NPGA to the problem from 8.32, while Figure 141 shows the application of NGPA to the problem defined in equations 8.71-8.72. In both cases, NPGA is able to find several points near the Pareto frontier for each problem.

Niched Pareto Genetic Algorithm

I-1   Independent variables $\vec{x}$

       Objective function

I-2

$$\vec{F}(\vec{x}) = \begin{bmatrix} f_1(\vec{x}) \\ f_2(\vec{x}) \\ \vdots \\ f_n(\vec{x}) \end{bmatrix}$$

I-3   Population size N

I-4   Minimum sharing distance σ_{min}

S-1   Create an initial population of N states chosen at random

S-2   Set the fitness of each state to zero

       For each state, count the number of states where $d < \sigma_{min}$. For each state where this condition add a factor of

S-3

$$S = 1 - \frac{d}{\sigma_{min}}$$

       to the fitness for the state.

S-4   Apply tournament selection with replacement, mutation, and recombination to generate a new population based on the current population

S-5   Add the new population to the archive of nondominated states. Update the archive to maintain nondominance among the states.

S-6   Set the population to the new population

S-7   Repeat from S-2 for as many iterations as desired

O-1   The archive represents the approximation to the Pareto frontier

Algorithm XLVI: NPGA is an evolutionary based algorithm that may be used to identify the Pareto frontier.

8.17   Weighted Metric Method

The weighted metric method optimizes a general distance between a reference point and the objective function. The metric is parameterized in terms of a set of weights and these weights varied. A solution is obtained for each set of weights, and each of these solutions is on the boundary of the objective function. Nondominated points are retained to create an estimate to the Pareto frontier.

Let the multiobjective function be given by

$$\vec{F}(\vec{x}) = \begin{bmatrix} f_1(\vec{x}) \\ f_2(\vec{x}) \\ \vdots \\ f_n(\vec{x}) \end{bmatrix} \qquad 8.114$$

A reference point is chosen and the distance between the objective function and the reference point is optimized. Let $\vec{z}$ be the reference point. The generalized distance is given by

$$D(\vec{x}) = \left(\sum_{i=1}^{n} w_i [f_i(\vec{x}) - z_i]^p \right)^{1/p} \qquad 8.115$$

For example, if $p = 2$ and the weight vector $\vec{w} = \vec{1}$ we have the Euclidean distance

$$D(\vec{x}) = \left(\sum_{i=1}^{n} [f_i(\vec{x}) - z_i]^2 \right)^{1/2} \qquad 8.116$$

By optimizing this distance, a point on the boundary of the objective function is identified. As the weight vector is varied, emphasis is placed on one component over another, and a different point on the objective optimizes the distance.

As the value of p becomes large, the largest term in the sum dominates. As p approaches infinity, optimizing the generalized distance reduces to the Tchbycheff problem of either minimizing

$$max[w_1|f_1(\vec{x}) - z_1|, w_1|f_1(\vec{x}) - z_1|, ..., w_n|f_n(\vec{x}) - z_n|] \qquad 8.117$$

or maximizing

$$min[w_1|f_1(\vec{x}) - z_1|, w_1|f_1(\vec{x}) - z_1|, ..., w_n|f_n(\vec{x}) - z_n|] \qquad 8.118$$

If the reference point is placed near the Nadir point, then the distance is maximized. If the reference point is placed near the Utopia point, the distance is maximized.

Weighted Metric Method

I-1   Independent variables $\vec{x}$

Objective function

I-2
$$\vec{F}(\vec{x}) = \begin{bmatrix} f_1(\vec{x}) \\ f_2(\vec{x}) \\ \vdots \\ f_n(\vec{x}) \end{bmatrix}$$

I-3   Method for optimizing a single objective function $g(\vec{x})$

I-4   Reference point $\vec{z}$

I-5   Value of p for the metric

Chose a point $\vec{w}$ such that

S-1
$$\sum_{i=1}^{n} w_i = 1 \quad |w| \le 1$$

Maximize the single objective function

S-2
$$g(\vec{x}) = \left(\sum_{i=1}^{n} w_i [f_i(\vec{x}) - z_i]^p \right)^{1/p}$$

Let $\vec{x}^*$ designate this maximum.

S-3   Compute the value of $\vec{F}(\vec{x}^*)$

S-4   Add $\vec{F}(\vec{x}^*)$ to the list of potential points on the Pareto frontier and continue from S-1 for the desired number of $\vec{w}$ to compute

S-5   Iterate through the list of potential points for the Pareto frontier

S-6   Consider the ith point on the list $\vec{F}(\vec{x}_i^*)$

S-7   Compare $\vec{F}(\vec{x}_i^*)$ to each other point on the list of potential frontier points. If $\vec{F}(\vec{x}_i^*)$ is dominated by another point $\vec{F}(\vec{x}_j^*)$, eliminate $\vec{F}(\vec{x}_i^*)$ from the list. If $\vec{F}(\vec{x}_i^*)$ dominates any other point $\vec{F}(\vec{x}_k^*)$, eliminate $\vec{F}(\vec{x}_k^*)$ from the list.

O-1   The list of frontier points $\vec{F}(\vec{x}_i^*)$ remaining after S-7 is the approximation to the Pareto frontier.

Algorithm XLVII: The weighted metric method may be used to identify the Pareto frontier.

8.18 ε–Constraint Method

The ε-constraint method optimizes a multiobjective by optimizing a single component of the objective function and constraining the other components. By varying the constraints, different points on the boundary of the objective are discovered.

Let the multiobjective function be given by

$$\vec{F}(\vec{x}) = \begin{bmatrix} f_1(\vec{x}) \\ f_2(\vec{x}) \\ \vdots \\ f_n(\vec{x}) \end{bmatrix} \qquad\qquad 8.119$$

Let c be the component of the objective selected for optimization. The redefined multiobjective problem is

$$\vec{F}(\vec{x}) = f_c(\vec{x})$$
$$such\ that \qquad\qquad 8.120$$
$$f_i(\vec{x}) \geq \varepsilon_i \quad i \neq c$$

The redefined problem is a constrained single objective maximization problem. The value $\vec{x}$ where the maximum occurs is recorded. The corresponding value $\vec{F}(\vec{x})$ is a point on the boundary of the objective.

An archive of nondominated points is updated with this boundary point, and the archive is adjusted to maintain a set of nondominated states. This archive is the approximation to the Pareto frontier.

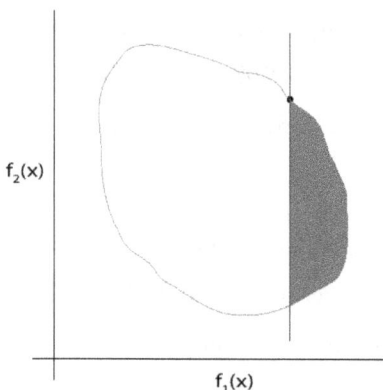

Figure 142: Example of application of constraints. The shaded region is a region where the first component is greater than or equal to a particular value. The point is the maximum of the second component under this constraint.

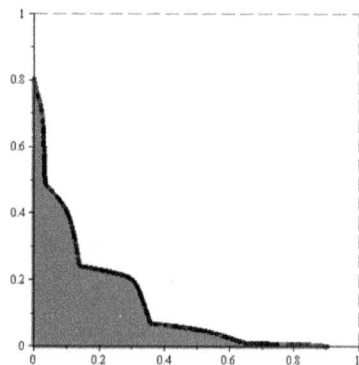

Figure 143: ε-constraint method applied to the problem from 8.71-8.72.

ε-Constraint Method

I-1   Independent variables $\vec{x}$

         Objective function

I-2

$$\vec{F}(\vec{x}) = \begin{bmatrix} f_1(\vec{x}) \\ f_2(\vec{x}) \\ \vdots \\ f_n(\vec{x}) \end{bmatrix}$$

I-3   Method for optimizing a single objective function $g(\vec{x})$

I-4   Component to optimize c

S-1   Use the single objective method to maximize each component of the objective function individually. Set μ_i as the maximum component value for the ith component.

S-2   Use the single objective method to minimize each component of the objective function individually. Set ν_i as the minimum component value for the ith component.

S-3   Choose a set of parameters ε_i where $\nu_i \leq \varepsilon_i \leq \mu_i$

         Maximize the single objective function

$$g(\vec{x}) = f_c(\vec{x})$$

S-4   subject to the constraints

$$f_i(\vec{x}) \geq \varepsilon_i \quad i \neq c$$

         Let $\vec{x}^*$ designate this maximum.

S-5   Compute the value of $\vec{F}(\vec{x}^*)$

S-6   Compare $\vec{F}(\vec{x}_i^*)$ to each other point on the archive of potential frontier points and adjust the archive to maintain a list of nondominated states

S-7   Repeat from S-3 as many times as desired

O-1   The archive is the approximation to the Pareto frontier

Algorithm XLVIII: The ε-constraint method may be used to identify the Pareto frontier.

8.19   Keeney-Raiffa Method

The Kenney-Raiffa method is a method for multiobjective optimization similar to the weighted sum method. In the weighted sum method, the multiobjective function is replaced by a single objective function where the components are summed together as a weighted sum.

The Kenney-Raiffa method multiplies the components together as a weighted product. The resulting produce is optimized to identify a point on the boundary, and the boundary points are compared to identify an approximation to the Pareto frontier.

Let the multiobjective function be given by

$$\vec{F}(\vec{x}) = \begin{bmatrix} f_1(\vec{x}) \\ f_2(\vec{x}) \\ \vdots \\ f_n(\vec{x}) \end{bmatrix} \qquad 8.121$$

The Keeney-Raffia method examines the weighted product

$$g(\vec{x}) = \prod_{k=1}^{n} [1 + w_i f_i(\vec{x})] \qquad 8.122$$

The method proceeds in iterations by choosing a weight vector $\vec{w}$ and optimizing $g(\vec{x})$. The optimum for a particular weight vector is a point on the boundary of the objective.

An archive of nondominated points is maintained during the execution of the algorithm. As a new boundary point is identified, the new point is compared with the archive.

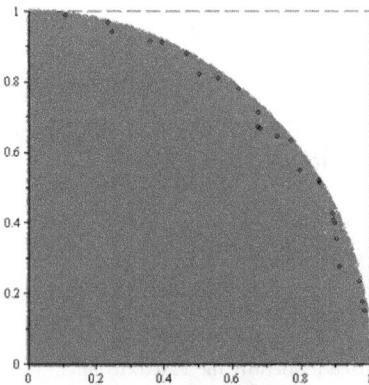

Figure 144: Keeney-Raiffa method applied to the quarter circle problem from 8.32.

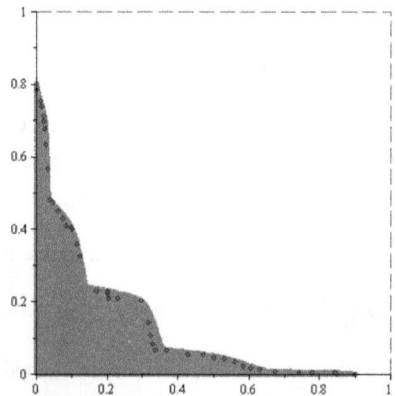

Figure 145: Keeney-Raiffa method applied to the problem from 8.71-8.72.

If the new point is dominated by any point in the archive, the point is discarded. If the new point is not dominated by any point in the archive, the new point is added to the archive, and the other points in the archive are examined.

Any archive point that is dominated by the new point is removed from the archive. In this manner the archive maintains a set of nondominated points, and these points are the approximation to the Pareto frontier.

Keeney-Raiffa Method

I-1   Independent variables $\vec{x}$

Objective function

I-2

$$\vec{F}(\vec{x}) = \begin{bmatrix} f_1(\vec{x}) \\ f_2(\vec{x}) \\ \vdots \\ f_n(\vec{x}) \end{bmatrix}$$

I-3   Method for optimizing a single objective function $g(\vec{x})$

Chose a point $\vec{w}$ such that

S-1

$$\sum_{i=1}^{n} w_i = 1 \quad |w| \leq 1$$

Maximize the single objective function

S-2

$$g(\vec{x}) = \prod_{k=1}^{n} [1 + w_i f_i(\vec{x})]$$

Let $\vec{x}^*$ designate this maximum.

S-3   Compute the value of $\vec{F}(\vec{x}^*)$

S-4   Compare $\vec{F}(\vec{x}_i^*)$ to each other point on the list of potential frontier points. If $\vec{F}(\vec{x}_i^*)$ is dominated by another point $\vec{F}(\vec{x}_j^*)$, eliminate $\vec{F}(\vec{x}_i^*)$ from the list. If $\vec{F}(\vec{x}_i^*)$ dominates any other point $\vec{F}(\vec{x}_k^*)$, eliminate $\vec{F}(\vec{x}_k^*)$ from the list.

O-1   The list of frontier points $\vec{F}(\vec{x}_i^*)$ remaining after S-7 is the approximation to the Pareto frontier.

Algorithm XLIX: The Keeney-Raiffa method may be used to identify the Pareto frontier.

9 Multiobjective Performance Measures

9.1 Measuring Performance of Multiobjective Optimization

Measuring performance for multiobjective optimization problems is different than for single objective optimization. In single objective optimization, there is a unique optimum value for the objective. There may be multiple values for the parameters where the objective has this unique value, but there is a single optimum value for the objective.

Performance measures for single objective optimization problems typically examine three parameters. Each of these is reviewed below.

Efficiency

First is an efficiency measure that counts the number of times the objective is evaluated. In many cases, evaluating the objective function is computationally expensive. Counting the number of evaluations of the objective is indicative of the amount of computational time required.

Objective Distance

A second measure of performance in single objective optimization is the distance between the optimum objective value and the value achieved by the algorithm. This measures how close the solution is to the true value in objective space.

Parameter Distance

The third measure of performance is the distance between the optimum parameter value and the value found by the algorithm. This measure is similar to the previous measure except the distance is measured in the parameter space rather than the objective space.

Multiobjective optimization uses analogs of these performance measures as well as several additional measures. Since multiobjective optimization problems typically do not have a single optimum, measurements of distance to the optimum are extended to measurements of distance to the frontier. One difficulty particular to multiobjective optimization problems is that the true frontier may not be known. The frontier is often quite complicated, and an exact mathematical formulation may not be available.

Frontier Unknown

Efficiency

Efficiency is the count of the number of evaluations of the objective function. This is a direct analog of the efficiency for single objective optimization.

Component Efficiency

Component efficiency is the count of the number of component evaluations of the objective function required. Some multiobjective optimization techniques only evaluate a single component. By measuring efficiency in terms of component evaluations, these techniques are not unduly biased when comparing to techniques that evaluate all components.

Number of Nondominated Points

Multiobjective optimization problems often have solutions in terms of a frontier of points. Optimization algorithms may find multiple nondominated points as an approximation to this frontier. Counting the number of nondominated points is an estimate of the utility of the algorithm. The more nondominated points identified, the better the performance of the algorithm.

Average Minimum Distance between Nondominated Points

Application of a multiobjective optimization algorithm arrives at a set of nondominated points meant as an approximation to the Pareto frontier. One important measure is how well these points cover the frontier. This metric sums the distance between each point and its nearest neighbor, and averages the result.

Multiset Coverage

In some cases it is desirable to compare multiple algorithms against each other. The approach of this metric is to apply several multiobjective optimization algorithms and have each approximate the frontier. The results are combined together into a single frontier retaining only nondominated solutions. From this, the percentage of points in the final frontier that result from each algorithm is computed.

Frontier Known

Objective Distance to the Frontier

Measuring the distance between a set of points and the Pareto frontier may be computed in two varieties. One variant measures how close the points identified by the algorithm are to the frontier, while the other measures how close points on the frontier are to the points identified by the algorithm. The first is useful in determining if the points identified are truly on the frontier. The second is useful in measuring how well the points identify the features of the frontier.

Average Distance between Points Found and the Frontier

For each nondominated frontier point identified by the algorithm, compute the minimum distance between the point and the true frontier, then average the results. This metric is useful to determine how well the points measure against the true frontier.

Average Distance between the Frontier and Points Found

In this variant, several points on the frontier are selected, and the minimum distance between these points and the points obtained from the algorithm is computed and averaged. This metric is useful to measure how well the frontier is approximated by the points identified by the algorithm.

Parameter Distance to the Frontier

This metric is similar to the previous metric expect the measurements are carried out in the parameter space rather than the objective space. The measurements are carried out in essentially the same manner.

Average Distance between Points Found and the Frontier

For each nondominated point identified by the algorithm, compute the minimum distance between the associated point in parameter space and the region of the parameter space that corresponds to the frontier. This distance is computed for each nondominated point, and the results averaged.

Average Distance between the Frontier and Points Found

In parameter space, select several points that correspond to frontier points in the objective space. For each of these points, compute the minimum distance between the point and the points identified by the algorithm. Sum the minimum distances for each of the points selected, and average the results.

Number of Frontier Points Identified

This metric is a simple count of the number of points identified by the algorithm that are in fact on the frontier.

Count of Number of Points on the Frontier

If the multiobjective is discrete, then count the number of points identified by the algorithm that are exactly on the true frontier.

Count of Number of Nondominated Points within Tolerance

For continuous multiobjectives, it is unlikely that any point identified by the algorithm is exactly on the frontier. In this case, a tolerance interval is specified. Points identified by the algorithm that are within this tolerance are counted as points on the frontier, while points that are outside of the tolerance window are considered off the frontier.

Percent of Hypercells Occupied

In this metric, the frontier is divided up into a set of hypercells with each cell representing a portion of the frontier. Each of the nondominated points identified by the algorithm is assigned to zero or one of these hypercells.

If a point lies within a hypercell, it is assigned to that hypercell. However, it may be the case that the algorithm has identified points that are not close enough to the frontier to fall into a hypercell. Such points are not assigned to any of hypercell.

From these assignments, the number of hypercells that have at least one point is counted. This number is divided by the total number of hypercells to obtain an estimate of the percent coverage of the frontier represented by the points identified with the algorithm.

This method may be applied in either the objective or parameter spaces.

The above is a general list of the types of metrics useful in measuring the performance of multiobjective optimization algorithms. In order to compute any of these metrics, a test problem must be identified and the algorithms applied to the test problem.

Not surprisingly, there is no single metric that is useful for measuring the performance of a multiobjective optimization problem. The metrics above measure different aspects of performance, and each aspect may be of more or less importance in a particular situation.

9.2 Multiobjective Functions

3D-STAR				
Definition			**Objective**	**Frontier**
$f_1(\vec{x}) = .5 + \dfrac{z}{2\sqrt{2}}[\cos^3(2\pi y) - \sin^3(2\pi y)][\cos^3(2\pi x) + \sin^3(2\pi x)]$ $f_2(\vec{x}) = .5 + \dfrac{z}{2\sqrt{2}}[\cos^3(2\pi y) + \sin^3(2\pi y)][\cos^3(2\pi x) + \sin^3(2\pi x)]$ $f_3(\vec{x}) = .5 + \dfrac{z}{2\sqrt{2}}[\cos^3(2\pi y) - \sin^3(2\pi y)][\cos^3(2\pi x) - \sin^3(2\pi x)]$				
			Parameters	**Frontier**
Objectives	3	**Parameters**	3	
Continuous	N	**Range**	$0 \le x_i \le 1$	
Concave	N	**Isolated**	Y	

BINORMAL				

Definition			Objective	Frontier
$f_1(\vec{x}) = xy$ $f_2(\vec{x}) = \dfrac{y[\exp(-10(x - .25)^2) + .8\exp(-25(x - .75)^2)]}{1.001544375}$				

			Parameters	Frontier
Objectives	2	Parameters	2	
Continuous	Y	Range	$0 \le x_i \le 1$	
Concave	N	Isolated	N	

BOY'S SURFACE				

Definition			Objective	Frontier
$f_1(\vec{x}) = \dfrac{2}{3}\dfrac{\left[\cos(2\pi x)\cos(4\pi y) + \sqrt{2}\sin(2\pi x)\cos(2\pi y)\right]\cos(2\pi x)}{\sqrt{2} - \sin(4\pi x)\sin(6\pi y)}$ $f_2(\vec{x}) = \dfrac{2}{3}\dfrac{\left[\cos(2\pi x)\sin(4\pi y) - \sqrt{2}\sin(2\pi x)\sin(2\pi y)\right]\cos(2\pi x)}{\sqrt{2} - \sin(4\pi x)\sin(6\pi y)}$ $f_3(\vec{x}) = -\dfrac{\sqrt{2}\cos^2(2\pi x)}{\sqrt{2} - \sin(4\pi x)\sin(6\pi y)}$				

			Parameters	Frontier
Objectives	3	Parameters	2	
Continuous	Y	Range	$-1 \le x_i \le 1$	
Concave	N	Isolated	N	

BUTTERFLY				

Definition			Objective	Frontier
$f_1(\vec{x}) = .5 + \dfrac{x}{6}\cos(2\pi y)\left[1 + 2\cos^3(8\pi y) - \sin(4\pi y)\right]$ $f_2(\vec{x}) = .5 + \dfrac{x}{6}\sin(2\pi y)\left[1 + 2\cos^3(8\pi y) - \sin(4\pi y)\right]$				

			Parameters	Frontier
Objectives	2	Parameters	2	
Continuous	N	Range	$0 \le x_i \le 1$	
Concave	N	Isolated	Y	

CIRCLE

Definition	Objective	Frontier

$$f_1(\vec{x}) = x \cos\left(\frac{\pi}{2}y\right)$$

$$f_2(\vec{x}) = x \sin\left(\frac{\pi}{2}y\right)$$

		Parameters	Frontier

Objectives	2	Parameters	2
Continuous	Y	Range	$0 \leq x_i \leq 1$
Concave	N	Isolated	N

CLOVER

Definition	Objective	Frontier

$$f_1(\vec{x}) = .5 + \frac{x}{2}\cos(2\pi y)\left[\sin(4\pi y) + .25\sin(12\pi y)\right]$$

$$f_2(\vec{x}) = .5 + \frac{x}{2}\sin(2\pi y)\left[\sin(4\pi y) + .25\sin(12\pi y)\right]$$

		Parameters	Frontier

Objectives	2	Parameters	2
Continuous	Y	Range	$0 \leq x_i \leq 1$
Concave	Y	Isolated	N

DUMBBELL

Definition	Objective	Frontier

$$f_1(\vec{x}) = .5 + x\frac{\cos^3\left(2\pi(y + 0.125)\right)}{2}$$

$$f_2(\vec{x}) = .5 + x\frac{\sin^3(2\pi y)}{2}$$

		Parameters	Frontier

Objectives	2	Parameters	2
Continuous	Y	Range	$0 \leq x_i \leq 1$
Concave	Y	Isolated	N

EXPONENTIAL MULTI-MODAL

Definition	Objective	Frontier

$$f_1(\vec{x}) = x$$
$$f_2(\vec{x}) = e^{-\left(250(x-.2)^2\right)} + e^{-\left(2.5(x-.6)^2\right)}$$

		Parameters	Frontier

Objectives	2	Parameters	2
Continuous	Y	Range	$0 \le x_i \le 1$
Concave	N	Isolated	N

FIGURE EIGHT

Definition	Objective	Frontier

$$f_1(\vec{x}) = \frac{1}{2} + \frac{(2 + .2\sin(4\pi x))\sin(2\pi x)}{8}$$
$$f_2(\vec{x}) = \frac{1}{2} + \frac{3.2\cos(4\pi x)}{8}$$
$$f_3(\vec{x}) = \frac{1}{2} + \frac{(2 + .2\sin(4\pi x))\cos(2\pi x)}{8}$$

		Parameters	Frontier

Objectives	3	Parameters	1
Continuous	Y	Range	$0 \le x_i \le 1$
Concave	N	Isolated	N

HELIX

Definition	Objective	Frontier

$$f_1(\vec{x}) = x\frac{(1 + \cos(20\pi x))}{2}$$
$$f_2(\vec{x}) = x\frac{(1 + \sin(20\pi x))}{2}$$
$$f_3(\vec{x}) = x$$

		Parameters	Frontier

Objectives	3	Parameters	1
Continuous	Y	Range	$0 \le x_i \le 1$
Concave	N	Isolated	N

HYPER-SINE

Definition			Objective	Frontier

$$f_1(\vec{x}) = \frac{x(1-y)}{2}[1 + \sin(5\pi x)]$$

$$f_2(\vec{x}) = \frac{y(1-x)}{2}[1 + \cos(5\pi y)]$$

			Parameters	Frontier

Objectives	2	Parameters	2
Continuous	Y	Range	$0 \le x_i \le 1$
Concave	Y	Isolated	N

IRRATIONAL HYPERSINE

Definition			Objective	Frontier

$$f_1(\vec{x}) = \frac{\sqrt{x}(1-y)}{2}(1 + \cos(2000\pi y))$$

$$f_2(\vec{x}) = \frac{\sqrt{y}(1-x)}{2}\left(1 + \cos(2000\sqrt{2}\pi y)\right)$$

			Parameters	Frontier

Objectives	2	Parameters	2
Continuous	N	Range	$0 \le x_i \le 1$
Concave	Y	Isolated	Y

KLEIN BOTTLE

Definition			Objective	Frontier

$$p_z = \cos^2(\pi x)$$
$$p_y = -\cos^3(\pi x)\sin(\pi x)$$
$$t_z = -\cos(\pi x)[4\cos^2(\pi x) - 3][3\sin(2\pi x) + 4]/36$$
$$t_y = \sin(\pi x)[3\sin(\pi x) + 4]/18$$
$$t_x = \sqrt{t_z^2 + t_y^2}$$
$$f_1(\vec{x}) = t_x \sin(2\pi y)$$
$$f_2(\vec{x}) = p_y + t_y \cos(2\pi y)$$
$$f_3(\vec{x}) = p_z + t_z \cos(2\pi y)$$

			Parameters	Frontier

Objectives	3	Parameters	2
Continuous	Y	Range	$0 \le x_i \le 1$
Concave	N	Isolated	N

LINEAR SINE				
Definition			**Objective**	**Frontier**
$f_1(\vec{x}) = (1 - x)(.5 + .25\sin(10\pi x))$ $f_2(\vec{x}) = x$				
			Parameters	**Frontier**
Objectives	2	Parameters	1	
Continuous	N	Range	$0 \le x_i \le 1$	
Concave	N	Isolated	N	

OCTAHEDRON				
Definition			**Objective**	**Frontier**
$f_1(\vec{x}) = .5 + \dfrac{z}{2}\cos^3(2\pi y)\sin^3(2\pi x)$ $f_2(\vec{x}) = .5 + \dfrac{z}{2}\sin^3(2\pi y)\sin^3(2\pi x)$ $f_3(\vec{x}) = .5 + \dfrac{z}{2}\cos^3(2\pi x)$				
			Parameters	**Frontier**
Objectives	3	Parameters	3	
Continuous	Y	Range	$0 \le x_i \le 1$	
Concave	Y	Isolated	N	

STAR				
Definition			**Objective**	**Frontier**
$f_1(\vec{x}) = .5 + x\dfrac{\cos^3(2\pi y) - \sin^3(2\pi y)}{2\sqrt{2}}$ $f_2(\vec{x}) = .5 + x\dfrac{\cos^3(2\pi y) + \sin^3(2\pi y)}{2\sqrt{2}}$				
			Parameters	**Frontier**
Objectives	2	Parameters	2	
Continuous	N	Range	$0 \le x_i \le 1$	
Concave	N	Isolated	Y	

TREFOIL

Definition			Objective	Frontier

$$f_1(\vec{x}) = \frac{1}{2} + \frac{(2 + \cos(6\pi x))\cos(4\pi x)}{8}$$
$$f_2(\vec{x}) = \frac{1}{2} + \frac{(2 + \cos(6\pi x))\sin(4\pi x)}{8}$$
$$f_3(\vec{x}) = \frac{1}{2} + \frac{\sin(6\pi x)}{8}$$

			Parameters	Frontier

Objectives	3	Parameters	1
Continuous	N	Range	$0 \le x_i \le 1$
Concave	N	Isolated	N

TRIANGLE

Definition	Objective	Frontier

$$f_1(\vec{x}) = x$$
$$f_2(\vec{x}) = (1 - x)y$$

	Parameters	Frontier

Objectives	2	Parameters	2
Continuous	N	Range	$0 \le x_i \le 1$
Concave	Y	Isolated	Y

X

Definition	Objective	Frontier

$$f_1(\vec{x}) = .5 + x\frac{\cos^3(2\pi(y + 0.125))}{2}$$
$$f_2(\vec{x}) = .5 + x\frac{\sin^3(2\pi y)}{2}$$

	Parameters	Frontier

Objectives	2	Parameters	2
Continuous	Y	Range	$0 \le x_i \le 1$
Concave	N	Isolated	N

9.3 Testing Multiobjective Optimization Algorithms

As discussed earlier, there are a wide variety of performance measures that may be applied to multiobjective optimization algorithms. Each performance measure examines different aspects of the optimization algorithm.

In this section we apply one such performance measure to the multiobjective problems from the previous section. The performance metric used is the multiset coverage. Each algorithm under examination is allowed to execute a predetermined number of function evaluations, and each results in an approximation to the Pareto frontier. The frontiers from each algorithm are combined into a single nondominated frontier.

The metric under examination is the number of points from the combined frontier that arise from each algorithm. Again, this is only one of many possible metrics, and the results provided here are not meant to indicate superior performance of any algorithm. Rather, this analysis is designed to provide insight to the methodologies for examining algorithm performance.

The table on the following page provides some results. Under this metric, algorithms that used the utopia hyperplane such as NBI, NC, DSD, etc. identify far fewer points than algorithms that use a genetic algorithm search. These algorithms use a distribution of points over the utopia hyperplane to achieve a disperse set of points along the frontier.

This metric does not measure the dispersion. Genetic algorithm methods often identify many Pareto incomparable points clustered into small regions. Hence, this metric has high values for algorithms based on these methods.

Moreover, algorithms that depend on a single dimensional optimization are also disadvantaged with this metric. These methods require many evaluations of the objective in order to optimize the single objective. Hence, there are multiple evaluations of the objective to compute a single point on the frontier.

Furthermore, algorithms that optimize a single objective often have difficulty with objectives that are functions of a single parameter. In this case, the intersection between the single objective and the frontier is likely the only allowable point. Hence, these algorithms may not be able to identify many points along the frontier because of difficulties in matching the constraints.

Again, there is no single metric that may be employed to compare multiobjective optimization algorithms. For a specific problem, appropriate metrics should be used in order to identify algorithms with the desired characteristics.

	MC	WS	NBI	NC	SPEA2	NSGA2	DSD	RWGA	ULWS
LinearSine	3620	10	2	5	435	3392	5	3610	9
TriangleI	88	6	9	9	138	43	1	435	7
MultiSphere	19	10	12	11	105	7	2	165	10
MultiBiNormal	9	9	12	8	100	9	1	536	9
HyperSine	99	4	5	9	234	7	2	248	8
ExpMultiModal	4660	10	5	11	994	4317	4	4747	10
Butterfly	1	10	8	3	106	70	2	19	10
Clover	3	8	1	5	130	162	0	63	8
Star	0	0	0	0	0	0	0	0	0
Dumbbell	24	10	11	7	161	801	1	96	10
X	1	10	1	11	37	104	1	10	9
IrrHyperSine	15	5	5	3	418	7	0	154	5
Helix	464	7	0	0	460	416	1	482	7
Trefoil	1963	10	3	7	602	1819	3	1947	10
FigureEight	2155	10	3	5	540	2096	3	2279	3
BoysSurface	1859	10	10	11	716	1262	2	2087	4
KleinBottle	1002	10	5	10	355	662	4	1626	4
3dStar	0	0	0	1	0	0	0	0	0
Octahedron	1431	5	6	7	315	850	0	1305	2

	VEGA	PAES	NPGA	PESA2	DMOEA	WMM	eCM	KRM
LinearSine	3786	3504	3661	3654	3753	0	0	0
TriangleI	39	0	105	35	95	0	2	0
MultiSphere	12	0	42	6	37	0	2	0
MultiBiNormal	24	0	9	1	5	0	1	0
HyperSine	86	11	159	108	120	0	10	0
ExpMultiModal	4810	367	4761	4721	4672	0	1	0
Butterfly	2	0	12	5	2	0	1	0
Clover	1	0	6	1	3	0	1	0
Star	0	0	0	0	0	0	0	0
Dumbbell	9	23	16	9	42	0	2	0
X	0	0	1	0	0	0	1	0
IrrHyperSine	14	1	60	27	20	0	0	4
Helix	467	57	475	465	468	0	0	0
Trefoil	1941	880	2046	1993	1935	0	7	5
FigureEight	2285	654	2256	2174	2233	0	5	3
BoysSurface	1737	910	1683	1812	1745	0	10	10
KleinBottle	924	244	1085	1113	1119	1	11	10
3dStar	0	0	0	0	0	0	0	0
Octahedron	1130	137	932	829	964	0	3	0

Appendix A: Algorithms

Single Objective

Method of Stationary Points I

I-1   Dependent variable y

I-2   Independent variable x

I-3   Differentiable function $f(x)$ relating $y = f(x)$

I-4   Range of allowable values for x: $x_{min} \leq x \leq x_{max}$

S-1   Find the derivative $f'(x)$

S-2   Find $x_i^* = x | f'(x) = 0$ where $x_{min} \leq x \leq x_{max}$

S-3   Let N be the total number of points from S-**2**

S-4 $x_{g\_min} = \min\left(f(x_1^*), f(x_2^*), \ldots, f(x_N^*), f(x_{min}), f(x_{max})\right)$

S-5 $x_{g\_max} = \max\left(f(x_1^*), f(x_2^*), \ldots, f(x_N^*), f(x_{min}), f(x_{max})\right)$

O-1   Minimum: x_{g_min}

O-2   Maximum: x_{g_max}

Method of Stationary Points II

I-1 Dependent variable z

I-2 Independent variables $\vec{x}$

I-3 Differentiable function $f(\vec{x})$ relating $z = f(\vec{x})$

I-4 Region $\mathcal{R}$ over $\vec{x}$ to find optima

S-1 Find the partial derivatives $\dfrac{\partial f(\vec{x})}{\partial x_k}$

S-2 Find $\vec{x}_i^* = \vec{x}\,\Big|\,\dfrac{\partial f(\vec{x})}{\partial x_k} = 0\ \exists k$ where $\vec{x} \in \mathcal{R}$

S-3 Let N be the total number of points from S-2

S-4 $\vec{x}_{g\_min} = \min\!\left(f(\vec{x}_1^*), f(\vec{x}_2^*), \dots, f(\vec{x}_N^*), f(\mathcal{R})\right)$

S-5 $\vec{x}_{g\_max} = \max\!\left(f(\vec{x}_1^*), f(\vec{x}_2^*), \dots, f(\vec{x}_N^*), f(\mathcal{R})\right)$

O-1 Minimum: $\vec{x}_{g\_min}$

O-2 Maximum: $\vec{x}_{g\_max}$

Substitution

I-1 Dependent variable z

I-2 Independent variables $\vec{x}$

I-3 Differentiable function $f(\vec{x})$ relating $z = f(\vec{x})$

I-4 Constraint equation $g(\vec{x}) = 0$

S-1 Solve the constraint equation for one of the variables:
$$x_k = h(x_1, x_2, \dots, x_{k-1}, x_{k+1}, \dots)$$

S-2 Use S-1 to eliminate x_k from the objective function

S-3 Solve the resulting objective function using unconstrained techniques

Aggregation of Constraints

I-1 Dependent variable z

I-2 Independent variables $\vec{x}$

I-3 Differentiable function $f(\vec{x})$ relating $z = f(\vec{x})$

I-4 Equality constraint equations $g_k(\vec{x}) = 0$ where $k = 1 \dots k_{max}$

I-5 Equality constraint equations $h_l(\vec{x}) \leq 0$ where $l = 1 \dots l_{max}$

I-6 There must be at least one constraint. Thus,
$$k_{max} + l_{max} > 0$$

I-7 Single objective optimization method

S-1 Create a combined objective function as
$$L = f(\vec{x}) + \sum_{k=1}^{k_{max}} \lambda_k |g_k(\vec{x})| + \sum_{l=1}^{l_{max}} \eta_l \left| \max\left(0, h_l(\vec{x})\right) \right|$$

S-2 Apply the single optimization method to the function L defined in step S-1 above

O-1 Solutions from the optimization method

Lagrange Multipliers

I-1 Dependent variable z

I-2 Independent variables $\vec{x}$

I-3 Differentiable function $f(\vec{x})$ relating $z = f(\vec{x})$

I-4 Constraint equation $g(\vec{x}) = 0$

S-1 Create a combined objective function as
$$L = f(\vec{x}) + \lambda g(\vec{x})$$

S-2 Compute the partial derivatives of L with respect to variables and λ

S-3 Solve the resulting equations for the variables and λ

O-1 k solutions $(\vec{x}_k, \lambda_k)$

Newton's Method I

I-1 Dependent variable z

I-2 Independent variable x

I-3 Single variable, twice differentiable function $f(x)$ relating $z = f(x)$

I-4 Starting point to seed the algorithm x_0

I-5 Uncertainty tolerance ε

S-1 Compute the first derivative of the objective function $f'(x)$

S-3 Compute $x_{n+1} = x_n - \dfrac{f(x_n)}{f'(x_n)}$

S-4 Repeat S-3 until $|x_{n+1} - x_n| < \varepsilon$

O-1 x_{n+1} value of a zero for the objective function

Newton's Method II

I-1 Dependent variable z

I-2 Independent variables $\vec{x}$

I-3 Multivariable, twice differentiable function $f(\vec{x})$ relating $z = f(\vec{x})$

I-4 Starting point to seed the algorithm $\vec{x}_0$

I-5 Uncertainty tolerance ε

I-6 Step size γ

S-1 Compute the gradient of the objective function $\nabla f(\vec{x})$

S-2 Compute the Hessian of the objective function $Hf(\vec{x})$

S-3 Compute $\vec{x}_{n+1} = \vec{x}_n - \gamma [H^{-1} f(\vec{x}_n)] \nabla f(\vec{x}_n)$

S-4 Repeat S-3 until $|\vec{x}_{n+1} - \vec{x}_n| < \varepsilon$

O-1 $\vec{x}_{n+1}$ value of a stationary point the objective function

Steepest Descents

I-1   Dependent variable z

I-2   Independent variables $\vec{x}$

I-3   Multivariable, differentiable function $f(\vec{x})$ relating $z = f(\vec{x})$

I-4   Starting point to seed the algorithm $\vec{x}_0$

I-5   Uncertainty tolerance ε

I-6   Step size γ

S-1   Compute the gradient of the objective function $\nabla f(\vec{x})$

S-2   Compute $\vec{x}_{n+1} = \vec{x}_n - \gamma \nabla f(\vec{x}_n)$

S-3   Repeat S-2 until $|\vec{x}_{n+1} - \vec{x}_n| < \varepsilon$

O-1 $\vec{x}_{n+1}$ value of a stationary point for the objective function

Bisection Method

I-1   Dependent variable z

I-2   Independent variables x

I-3   Univariate, function $f(x)$ relating $z = f(x)$

I-4   Starting points that bracket exactly one zero of $f(x)$: x_L and x_R

I-5   Uncertainty tolerance ε

S-1   Compute $x_T = \frac{x_L + x_R}{2}$

S-2   Compute $f(x_T)$

S-3   If $f(x_T)$ and $f(x_L)$ have the same sign, then set $x_L = x_T$

S-4   If $f(x_T)$ and $f(x_R)$ have the same sign, then set $x_R = x_T$

S-5   If $f(x_T) = 0$ then output x_T

S-6   Repeat from S-1 until $|x_L - x_R| < \varepsilon$

O-1   Estimate for the zero of the objective function: $x = \frac{x_L + x_R}{2}$

Secant Method I

I-1 Dependent variable z

I-2 Independent variables x

I-3 Univariate function $f(x)$ relating $z = f(x)$

I-4 Starting points where the derivative is different: x_0 and x_1, $f(x_0) \neq f(x_1)$

I-5 Uncertainty tolerance ε

S-1 Compute $x_{n+1} = x_n - \dfrac{f(x_n)}{f(x_n)-f(x_{n-1})}(x_n - x_{n-1})$

S-2 Repeat from S-1 until $|x_n - x_{n-1}| < \varepsilon$

O-1 Estimate for the zero of the objective function: x_{n+1}

Secant Method II

I-1 Dependent variable z

I-2 Independent variables x

I-3 Univariate function $f(x)$ relating $z = f(x)$

I-4 Starting points where the derivative has opposite signs: x_L and x_R, $f(x_L)f(x_R) < 0$. x_L is to the left of the zero, while x_R is to the right of the zero.

I-5 Uncertainty tolerance ε

S-1 Compute $x_T = x_R - \dfrac{f(x_R)}{f(x_R)-f(x_L)}(x_R - x_L)$

S-2 If $f(x_T)$ and $f(x_L)$ have the same sign, then set $x_L = x_T$

S-3 If $f(x_T)$ and $f(x_R)$ have the same sign, then set $x_R = x_T$

S-4 Repeat from S-1 until $|x_n - x_{n-1}| < \varepsilon$

O-1 Estimate for the zero of the objective function: x_n

Dekker's Method

I-1   Dependent variable z

I-2   Independent variables x

I-3   Univariate function $f(x)$ relating $z = f(x)$

I-4   Starting points where the derivative has opposite signs: x_L and x_R, $f(x_L)f(x_R) < 0$. x_L is to the left of the zero, while x_R is to the right of the zero.

I-5   Uncertainty tolerance ε

S-1   Set $b_n = \begin{cases} x_L & |f(x_L)| < |f(x_R)| \\ x_R & |f(x_R)| < |f(x_L)| \end{cases}$

S-2   Set $a_n = \begin{cases} x_L & |f(x_L)| > |f(x_R)| \\ x_R & |f(x_R)| > |f(x_L)| \end{cases}$

S-3   Compute $S_T = b_n - \dfrac{f(b_n)}{f(b_n)-f(b_{n-1})}(b_n - b_{n-1})$

S-4   Compute $B_T = \dfrac{b_n+a_n}{2}$

S-5   Set $N_T = \begin{cases} S_T & min(b_n, B_T) < S_T < max(b_n, B_T) \\ B_T & otherwise \end{cases}$

S-6   If $f(N_T)$ and $f(x_L)$ have the same sign, then set $x_L = N_T$

S-3   If $f(N_T)$ and $f(x_R)$ have the same sign, then set $x_R = N_T$

S-4   Repeat from S-1 until $|b_n - b_{n-1}| < \varepsilon$

O-1   Estimate for the zero of the objective function: b_n

Brent's Method

I-1   Dependent variable z

I-2   Independent variables x

I-3   Univariate function $f(x)$ relating $z = f(x)$

I-4   Starting points where the derivative has opposite signs: x_L and x_R, $f(x_L)f(x_R) < 0$. x_L is to the left of the zero, while x_R is to the right of the zero.

I-5   Uncertainty tolerance ε

S-1   Set $b_n = \begin{cases} x_L & |f(x_L)| < |f(x_R)| \\ x_R & |f(x_R)| < |f(x_L)| \end{cases}$

S-2   Set $a_n = \begin{cases} x_L & |f(x_L)| > |f(x_R)| \\ x_R & |f(x_R)| > |f(x_L)| \end{cases}$

S-3   Compute $S_T = b_n - \dfrac{f(b_n)}{f(b_n)-f(b_{n-1})}(b_n - b_{n-1})$

S-4   Compute $B_T = \dfrac{b_n+a_n}{2}$

S-5   Set $N_T = \begin{cases} S_T & min(b_n, B_T) < S_T < max(b_n, B_T) \\ B_T & otherwise \end{cases}$

S-6   If $N_T = S_T$, and if the previous iteration used B_T, and either
$$|\delta| < |b_n - b_{n-1}|$$
$$|S_T - b_n| < \frac{|b_n - b_{n-1}|}{2}$$
are not satisfied, then set $N_T = S_T$.

S-7   If $N_T = S_T$, and if the previous iteration used S_T, and either
$$|\delta| < |b_{n-1} - b_{n-2}|$$
$$|S_T - b_{n-1}| < \frac{|b_{n-1} - b_{n-2}|}{2}$$
are not satisfied, then set $N_T = S_T$.

S-6   If $f(N_T)$ and $f(x_L)$ have the same sign, then set $x_L = N_T$

S-3   If $f(N_T)$ and $f(x_R)$ have the same sign, then set $x_R = N_T$

S-4   Repeat from S-1 until $|b_n - b_{n-1}| < \varepsilon$

O-1   Estimate for the zero of the objective function: b_n

Halley's Method

I-1   Dependent variable z

I-2   Independent variable x

I-3   Single variable, twice differentiable function $f(x)$ relating $z = f(x)$

I-4   Starting point to seed the algorithm x_0

I-5   Uncertainty tolerance ε

S-1   Compute the first derivative of the objective function $f'(x)$

S-2   Compute the second derivative of the objective function $f''(x)$

S-3   Compute $x_{n+1} = x_n - \dfrac{2f(x_n)f'(x_n)}{2[f'(x_n)]^2 - f(x_n)f''(x_n)}$

S-4   Repeat S-3 until $|x_{n+1} - x_n| < \varepsilon$

O-1 x_{n+1} value of a zero for the objective function

Householder's Method

I-1   Dependent variable z

I-2   Independent variable x

I-3   Single variable, k times differentiable function $f(x)$ relating $z = f(x)$

I-4   Starting point to seed the algorithm x_0

I-5   Uncertainty tolerance ε

S-1   Compute the first k derivatives of $1/f(x)$

S-3   Compute $x_{n+1} = x_n + k\,\dfrac{\frac{d^{k-1}}{dx^{k-1}}(1/f(x_n))}{\frac{d^k}{dx^k}(1/f(x_n))}$

S-4   Repeat S-3 until $|x_{n+1} - x_n| < \varepsilon$

O-1 x_{n+1} value of a zero for the objective function

Inverse Quadratic Interpolation

I-1   Dependent variable z

I-2   Independent variable x

I-3   Single variable function $f(x)$ relating $z = f(x)$

I-4   Three starting points to seed the algorithm x_0, x_1, and x_2.

I-5   Uncertainty tolerance ε

Compute

S-1
$$x_{n+1} = \frac{f_{n-1}f_n}{(f_{n-2} - f_{n-1})(f_{n-2} - f_n)}x_{n-2} + \frac{f_{n-2}f_n}{(f_{n-1} - f_{n-2})(f_{n-1} - f_n)}x_{n-1}$$
$$+ \frac{f_{n-2}f_{n-1}}{(f_n - f_{n-2})(f_n - f_{n-1})}x_n$$

S-2   Repeat S-1 until $|x_{n+1} - x_n| < \varepsilon$

O-1 x_{n+1} value of a zero for the objective function

Müller's Method

I-1   Dependent variable z

I-2   Independent variable x

I-3   Single variable function $f(x)$ relating $z = f(x)$

I-4   Three starting points to seed the algorithm x_0, x_1, and x_2.

I-5   Uncertainty tolerance ε

S-1   Compute $w = \dfrac{f_n - f_{n-1}}{x_n - x_{n-1}} + \dfrac{f_n - f_{n-2}}{x_n - x_{n-2}} - \dfrac{f_{n-1} - f_{n-2}}{x_{n-1} - x_{n-2}}$

S-2   Compute $h = \dfrac{\frac{f_n - f_{n-1}}{x_n - x_{n-1}} - \frac{f_{n-1} - f_{n-2}}{x_{n-1} - x_{n-2}}}{x_n - x_{n-2}}$

S-3   Compute $x_{n+1} = x_n - \dfrac{2f(x_n)}{w \pm \sqrt{w^2 - 4f(x_n)h}}$

S-4   Repeat from S-1 until $|x_{n+1} - x_n| < \varepsilon$

O-1 x_{n+1} value of a zero for the objective function

Steffensen's Method

I-1   Dependent variable z

I-2   Independent variable x

I-3   Single variable function $f(x)$ relating $z = f(x)$

I-4   Starting points to seed the algorithm x_0.

I-5   Uncertainty tolerance ε

S-1   Compute $x_{n+1} = x_n - \dfrac{f^2(x_n)}{f(x_n + f(x_n)) - f(x_n)}$

S-3   Repeat from S-1 until $|x_{n+1} - x_n| < \varepsilon$

O-1 x_{n+1} value of a zero for the objective function

Evolutionary Algorithms

I-1   Dependent variable z

I-2   Independent variables $\vec{x}$

I-3   Objective function $f(\vec{x})$ relating $z = f(\vec{x})$

I-4 N initial points $\vec{x}^0{}_n$

S-1   Evaluate the objective function at each of the N points $\vec{x}^k{}_n$

S-2   Identify a set of M test points based on $\vec{x}^k{}_n$

S-3   Evaluate the objective function at each of the M test points

S-4   Use the evolutionary method to select L points for the next iteration

S-5   Repeat from S-1 as many times as desired

O-1   Each $\vec{x}^{n+1}{}_l$ represents a potential optima for the objective function

Relaxation

I-1   Dependent variable z

I-2   Independent variables $\vec{x}$

I-3   Differentiable function $f(\vec{x})$ relating $z = f(\vec{x})$

I-4   Initial point $\vec{x}^0$

I-5   Uncertainty tolerance ε

S-1   Find the partial derivatives $\dfrac{\partial f(\vec{x})}{\partial x_k}$

S-2   Find $\dfrac{\partial f(\vec{x})}{\partial x_k}$

S-3   Choose a random ordering for minimizing each component

Find the vector $\left(x_1^n, x_2^n, \ldots, x_l^{n+1}, \ldots, x_k^n\right)$ where

S-4
$$\frac{\partial f\left(\vec{x} = \left(x_1^n, x_2^n, \ldots, x_l^{n+1}, \ldots, x_k^n\right)\right)}{\partial x_k} = 0$$

and l is the first component to minimize.

Find the vector $\left(x_1^n, x_2^n, \ldots, x_m^{n+1}, \ldots, x_l^{n+1}, \ldots, x_k^n\right)$ where

S-5
$$\frac{\partial f\left(\vec{x} = \left(x_1^n, x_2^n, \ldots, x_m^{n+1}, \ldots, x_l^{n+1}, \ldots, x_k^n\right)\right)}{\partial x_k} = 0$$

and m is the second component to minimize.

S-6   Repeat the previous step for each component in the randomly chosen order to get the vector for the next iteration
$\vec{x}^{n+1} = \left(x_1^{n+1}, x_2^{n+1}, \ldots, x_k^{n+1}\right).$

S-7   Repeat from S-1 until $|\vec{x}^{n+1} - \vec{x}^n| < \varepsilon$

O-1 $\vec{x}^{n+1}$ value of a stationary point for the objective function

Genetic Algorithms

I-1   Dependent variable z

I-2   Independent variables $\vec{x}$

I-3   Identify the genes and a method for mapping gene sequences to vectors in the search space

I-4   Objective function $f(\vec{x})$ relating $z = f(\vec{x})$ where $\vec{x}$ is a sequence of genes (a chromosome)

I-5 N initial genes $\vec{x}^0{}_n$

I-6   A weighted random selection method for randomly choosing a chromosome from the population pool, where the weight of the random selection is dependent on the value of the objective function

S-1   Evaluate the objective function at each of the N points $\vec{x}^k{}_n$

S-2   Randomly select two chromosomes from the population pool. The random selection process is weighted by the value of the objective function.

S-3   Choose crossover or mutation. Variants of the algorithm include only using crossover or only using mutation.

S-4   Perform crossover or mutation on the two chromosomes randomly selected.

S-5   Place the two new chromosomes into the population pool for the next generation.

S-6   Repeat from S-2 until the desired number of chromosomes for the next generation population pool is achieved.

S-7   Evaluate the objective function for each chromosome in the next generation population pool.

S-8   If the best fitness for the next generation population pool is less than the best fitness from the current population pool, randomly select one of the chromosomes from the next generation and replace it with the best chromosome from the current generation.

S-9   Repeat from S-2 as many times as desired.

O-1   The best fitness from the current generation is potential optimum.

Quantum Particle Search

I-1   Dependent variable z

I-2   Independent variables $\vec{x}$

I-3   Objective function $f(\vec{x})$ relating $z = f(\vec{x})$

I-4 N initial states x_i

I-5   Value α used in the Laplace probability density

S-1   Iterate over each state $\vec{x}_i$

S-2   Compute $f(\vec{x}_i)$ for each state

S-3   Generate a random vector $\vec{r}$ where the components of $\vec{r}$ are generated according to the Laplace density
$$P(x) = \frac{\alpha}{2}exp(-\alpha|x|)$$

S-4   Compute
$$\vec{t} = \vec{x} + \vec{r}$$

S-5   If $f(\vec{t})$ is better than $f(\vec{x}_i)$, then replace $\vec{x}_i$ with $\vec{t}$ and process the next state

O-1   The best value of $f(\vec{x}_i)$ is the potential optimum.

Differential Evolution

I-1 Dependent variable z

I-2 Independent variables $\vec{x}$

I-3 Objective function $f(\vec{x})$ relating $z = f(\vec{x})$

I-4 N initial states $x^k{}_n$ (upper index is the state number, lower index designates the vector component)

I-5 Value for F

I-6 Value for CP

S-1 Evaluate the objective function at each of the N points $\vec{x}^k{}_n$

S-2 Iterate over each state vector $x^k{}_i$

S-3 Choose three distinct random states different from each other and $x^k{}_i$. Designate these as $x^a{}_i, x^b{}_i$, and $x^c{}_i$

S-4 Select a random vector component r

S-5 Construct a new state y_i from $x^k{}_i$ and $x^a{}_i, x^b{}_i$, and $x^c{}_i$ by examining each component:

S-6 Set $y_r = x^a{}_r + F\left(x^b{}_i - x^c{}_i\right)$

For the components $j \neq r$:

Choose a random variable R. If $R < CP$ then set

S-7
$$y_j = x^a{}_j + F\left(x^b{}_j - x^c{}_j\right)$$

otherwise set

$$y_j = x^k{}_j$$

S-8 If $f(y_r)$ is better than $f(x^k{}_i)$, then replace $x^k{}_i$ with y_r in the next generation. Otherwise, use $x^k{}_i$ in the next generation.

S-9 Repeat from S-2 for each state vector.

S-10 Repeat from S-1 for each iteration

O-1 The best fitness from the current generation is potential optimum.

Particle Swarm Optimization

I-1   Dependent variable z

I-2   Independent variables $\vec{x}$

I-3   Objective function $f(\vec{x})$ relating $z = f(\vec{x})$

I-4 N initial states $x^k{}_n$ randomly chosen in the search space

I-5   Initial velocity $v^k{}_n$ randomly chosen for each state

I-6   Value for ω

I-7   Value for ρ

I-8   Value for γ

I-9   Set the best known position for each particle to the initial position
$$p^k{}_i = x^k{}_i$$

I-10   Evaluate the objective function at each of the N points $x^k{}_n$

I-11   Set the best swarm position $\vec{s}_i$ to the state vector that has the best value for the objective function

S-1   For each state vector $x^k{}_i$:

S-2   Choose two random variables, R_1 and R_2, on the range $[0,1]$.

S-3   Update the velocity of the state by
$$v^k{}_i = \omega v^k{}_i + \rho R_1\left(p^k{}_i - x^k{}_i\right) + \gamma R_2\left(s_i - x^k{}_i\right)$$

S-4   Update the position of the state by
$$x^k{}_i = x^k{}_i + v^k{}_i$$

S-5   Compute the value of the objective function for this state $f\left(x^k{}_i\right)$

S-6   If $f\left(x^k{}_i\right)$ is better than $f\left(p^k{}_i\right)$ then set $p^k{}_i = x^k{}_i$

S-7   If $f\left(x^k{}_i\right)$ is better than $f(s_i)$ then set $s_i = x^k{}_i$

S-8   Repeat from S-1 for each state vector.

O-1 s_i is the potential optimum.

Amoeba Method

I-1   Dependent variable z

I-2   Independent variables $\vec{x}$

I-3   Objective function $f(\vec{x})$ relating $z = f(\vec{x})$

I-4 $N + 1$ initial vertices $\vec{v}_i$

I-5   Value for α

I-6   Value for r

I-7   Value for ρ

I-8   Value for σ

S-1   Compute $f(\vec{v}_i)$ for each vertex

S-2   Order the vertices according to their corresponding values for the objective functions, with the best value being the first vertex, second best as the second vertex, etc.

Compute

S-3
$$\vec{\mu} = \frac{1}{N} \sum_{i=1}^{N} \vec{v}_i$$

Note the sum does not include the worst vertex.

Compute

S-4
$$\vec{x}_s = \vec{\mu} + \alpha(\vec{\mu} - \vec{v}_{N+1})$$

and the corresponding value for the objective function.

S-5   If $f(\vec{x}_s)$ is better than the second worst vertex, but not better than the best vertex, replace the worst vertex with $\vec{x}_s$ and repeat from S-1.

If $f(\vec{x}_s)$ is better than the best vertex, then compute

$$\vec{x}_e = \vec{\mu} + r(\vec{\mu} - \vec{v}_{N+1})$$

S-6   If $f(\vec{x}_e)$ is better than $f(\vec{x}_s)$, then replace the worst vertex with $\vec{x}_e$ and repeat from S-1.

Otherwise, replace the worst vertex with $\vec{x}_s$ and repeat from S-1.

If $f(\vec{x}_s)$ is not better than the second worst vertex, compute

S-7
$$\vec{x}_c = \vec{v}_{N+1} + \rho(\vec{\mu} - \vec{v}_{N+1})$$

If $f(\vec{x}_c)$ is better than the worst vertex, then replace the worst vertex with $\vec{x}_c$ and repeat from S-1.

Replace all vertices except the best vertex using the formula

S-8
$$\vec{v}_i = \vec{v}_1 + \sigma(\vec{v}_i - \vec{v}_1)$$

Repeat from S-1.

O-1 $\vec{v}_1$ is the potential optimum.

Simulated Annealing

I-1   Dependent variable z

I-2   Independent variables $\vec{x}$

I-3   Objective function $f(\vec{x})$ relating $z = f(\vec{x})$

I-4 N initial states x_i

I-5   Schedule for the value of the temperature at each iteration.

S-1   Iterate over each state $\vec{x}_i$

S-2   Compute $f(\vec{x}_i)$ for each state

S-3   Generate a random test state, $\vec{r}_i$, based on $\vec{x}_i$

S-4   Compute

$$\Delta = -|f(\vec{r}_i) - f(\vec{r}_i)|$$

S-5   If $f(\vec{r}_i)$ is better than $f(\vec{x}_i)$, then replace $\vec{x}_i$ with $\vec{r}_i$ and process the next state

S-6   Otherwise compute the transition probability

$$P(\Delta, T) = e^{-\Delta/T}$$

S-7   Select a random variable $R \in [0,1)$

S-8   If $R < P(\Delta, T)$ then replace $\vec{x}_i$ with $\vec{r}_i$. Process the next iteration.

O-1   The best value of $f(\vec{x}_i)$ is the potential optimum.

Quadratic Vector Interpolation

I-1   Dependent variable z

I-2   Independent variables $\vec{x}$

I-3   Objective function $f(\vec{x})$ relating $z = f(\vec{x})$

I-4 N initial states x_i

I-5   Method for choosing uniform random deviates on the range $[0,1)$

S-1   Iterate over each state $\vec{x}_i$

S-2   Compute $f(\vec{x}_i)$ for each state

        Choose a random state $\vec{x}_j$ where $i \neq j$

S-3   Alternatively, choose three state vectors and compute

$$\vec{M} = \frac{1}{3}(\vec{x}_1 + \vec{x}_2 + \vec{x}_3)$$

S-4   Select a random $r \in [0,1)$

S-5   Compute $\vec{\sigma} = \vec{x}_j + r\vec{d}$

S-6   Compute $a = f(\vec{x}_j)$

S-7   Compute $b = \dfrac{\left[r^2\left(f(\vec{x}_i) - f(\vec{x}_j)\right) - \left(f(\vec{\sigma}) - f(\vec{x}_j)\right)\right]}{r(r-1)}$

S-8   Compute $c = \dfrac{\left[-r\left(f(\vec{x}_i) - f(\vec{x}_j)\right) + \left(f(\vec{\sigma}) - f(\vec{x}_j)\right)\right]}{r(r-1)}$

S-9   Compute $t = -\dfrac{b}{2c}$

S-10   Compute $\vec{p} = \vec{x}_j + t\vec{d}$

S-11   If $f(\vec{p})$ is better than $f(\vec{x}_i)$, replace $\vec{x}_i$ with $\vec{p}$ and process the next state

S-12   If $f(\vec{\sigma})$ is better than $f(\vec{x}_i)$, replace $\vec{x}_i$ with $\vec{\sigma}$ and process the next state

S-13   Repeat from S-3 as desired

O-1   The best value of $f(\vec{x}_i)$ is the potential optimum.

m-Dimensional Vector Interpolation

I-1   Dependent variable z

I-2   Independent variables $\vec{x}$

I-3   Objective function $f(\vec{x})$ relating $z = f(\vec{x})$

I-4 N initial states x_i

I-5   Value of the fit space m where m is less than or equal to the dimension of the search space

I-6   Specify the value σ

S-1   Select $m + 1$ vectors from the population $\vec{x}^i$

S-2   Compute $f(\vec{x}^i)$ for each state

S-3   Compute the difference vectors
$$\vec{\Delta}^i = \vec{x}^i - \vec{x}^0$$

S-4   Compute $\vec{\rho}$ as a m-dimensional vector whose components are uniform random deviates

S-5   Compute the vector $\vec{r}^i = \vec{x}^0 + \vec{\rho} \cdot \vec{\Delta}$

S-6   Compute the value of the objective function $f(\vec{r}^i)$

S-7   Repeat from S-4 to generate $\dfrac{m(m+1)}{2}$ vectors

S-8   Compute
$$\vec{d} = \begin{pmatrix} f(\vec{x}^1) - f(\vec{x}^0) \\ f(\vec{x}^2) - f(\vec{x}^0) \\ \vdots \\ f(\vec{x}^m) - f(\vec{x}^0) \\ f(\vec{r}^1) - f(\vec{x}^0) \\ f(\vec{r}^2) - f(\vec{x}^0) \\ \vdots \\ f\left(\vec{r}^{\frac{m(m+1)}{2}}\right) - f(\vec{x}^0) \end{pmatrix}$$

S-9   Compute

$$C = \begin{pmatrix} 1 & 0 & \cdots & 0 & 1 & 0 & \cdots & 0 \\ 0 & 1 & \cdots & 0 & 0 & 1 & \cdots & 0 \\ \vdots & \vdots & \vdots & \vdots & \vdots & \vdots & \vdots & \vdots \\ \rho_1^1 & \rho_2^1 & \cdots & \rho_m^1 & \rho_1^1\rho_1^1 & \rho_2^1\rho_1^1 & \cdots & \rho_m^1\rho_m^1 \\ \vdots & \vdots & \vdots & \vdots & \vdots & \vdots & \vdots & \vdots \\ \rho_1^m & \rho_2^m & \cdots & \rho_m^m & \rho_1^m\rho_1^m & \rho_2^m\rho_1^m & \cdots & \rho_m^m\rho_m^m \end{pmatrix}$$

S-10   Compute $\vec{p} = C^{-1}\vec{d}$

S-11   Compute $\vec{b}$ as the first m components of $\vec{p}$

S-12   Compute
$$\mathcal{A} = \begin{pmatrix} p_0p_0 & p_0p_1 & \cdots & p_0p_m \\ p_1p_0 & p_1p_1 & \cdots & p_1p_m \\ \vdots & \vdots & \vdots & \vdots \\ p_mp_0 & p_mp_1 & \cdots & p_mp_m \end{pmatrix}$$

S-13   Compute
$$\vec{t}^* = -\mathcal{A}^{-1}\vec{b}$$

S-14   Compute
$$\vec{x}^* = \vec{x}^0 + \vec{t}^* \cdot \vec{\Delta}$$

S-15   Compute $f(\vec{x}^*)$

S-16   Identify the vector in the set $\vec{x}^*$, $\vec{r}^l$ that has the best value for the objective function. Designate this vector as $\vec{x}'$

S-17   Identify the vector in the set $\vec{x}^k$ that has the worst value for the objective function. Designate this vector as $\vec{x}^w$

S-18   If $f(\vec{x}')$ is better than $f(\vec{x}^w)$, replace $\vec{x}^w$ with $\vec{x}'$, repeat from S-1.

S-19   Identify the vector in the set $\vec{x}^k$ that has the best value for the objective function. Designate this vector as $\vec{x}^b$

S-20   Replace all vectors $\vec{x}^k$ except $\vec{x}^b$ using the formula
$$\vec{x}^k = \vec{x}^b + \sigma(\vec{x}^k - \vec{x}^b)$$

S-21   Repeat from S-1

O-1   The best value of $f(\vec{x}_i)$ is the potential optimum.

Multiple Objectives

Turning Point Method

I-1   Independent variable x

Objective function

I-2
$$\vec{F}(x) = \begin{bmatrix} f_1(x) \\ f_2(x) \end{bmatrix}$$

S-1   Compute all points x where $\dfrac{\partial f_1(x)}{\partial x} = 0$ and $\dfrac{\partial^2 f_1(x)}{\partial x^2} \neq 0$ (the point x_i is either a minimum or maximum, not a point of inflection). Designate the set of all such points as the set X_1.

S-2   Compute all points x where $\dfrac{\partial f_2(x)}{\partial x} = 0$ and $\dfrac{\partial^2 f_2(x)}{\partial x^2} \neq 0$. Designate the set of all such points as the set X_2.

S-3   Compute the union of the sets $\bar{X} = X_1 \cup X_2$

S-4   Order the points $\bar{x}_i \in \bar{X}$ so that $\bar{x}_1 < \bar{x}_2 < \cdots < \bar{x}_p$

S-5   Compute $\vec{F}(\bar{x}_i)$ for each $\bar{x}_i \in \bar{X}$.

S-6   Identify each $\bar{x}_i$ where the corresponding $\vec{F}(\bar{x}_i)$ is not dominated by any other $\vec{F}(\bar{x}_j)$

S-7   Find the first $\bar{x}_i$ with a non-dominated $\vec{F}(\bar{x}_i)$

S-8   Examine $\bar{x}_{i+1}$:

  i.   If $\vec{F}(\bar{x}_i)$ dominates $\vec{F}(\bar{x}_{i+1})$, then none of the points $\vec{F}(x)$ where $\bar{x}_i \leq x \leq \bar{x}_{i+1}$ are on the Pareto frontier. Continue from S-9.

  ii. If $\vec{F}(\bar{x}_{i+1})$ is non-dominated, then all of the points $\vec{F}(x)$ where $\bar{x}_i \leq x \leq \bar{x}_{i+1}$ are on the Pareto frontier. Continue from S-9.

  iii. Otherwise, find the next sequential point $\bar{x}_j$ where $\vec{F}(\bar{x}_j)$ is non-dominated.

    e.   Case I:
$$f_1(x_i) > f_1(x_j)$$
$$f_2(x_i) < f_2(x_j)$$
    Find the minimum $x_b > x_i$ where $f_1(x_b) = f_1(x_j)$. All

points $\vec{F}(x)$ where $\bar{x}_i \leq x \leq \bar{x}_b$ are on the Pareto frontier. Continue from S-9.

f. Case II:

$$f_1(x_i) > f_1(x_j)$$
$$f_2(x_i) < f_2(x_j)$$

Find the minimum $x_b > x_i$ where $f_1(x_b) = f_1(x_j)$. All points $\vec{F}(x)$ where $\bar{x}_i \leq x \leq \bar{x}_b$ are on the Pareto frontier. Continue from S-9.

Examine $\bar{x}_{i-1}$:

iv. If $\vec{F}(\bar{x}_i)$ dominates $\vec{F}(\bar{x}_{i-1})$, then none of the points $\vec{F}(x)$ where $\bar{x}_{i-1} \leq x \leq \bar{x}_i$ are on the Pareto frontier. Continue from S-9.

v. If $\vec{F}(\bar{x}_{i+1})$ is non-dominated, then all of the points $\vec{F}(x)$ where $\bar{x}_{i-1} \leq x \leq \bar{x}_i$ are on the Pareto frontier. Continue from S-9.

vi. Otherwise, find the previous sequential point $\bar{x}_j$ where $\vec{F}(\bar{x}_j)$ is non-dominated.

g. Case I:

S-9

$$f_1(x_i) > f_1(x_j)$$
$$f_2(x_i) < f_2(x_j)$$

Find the minimum $x_b < x_i$ where $f_1(x_b) = f_1(x_j)$. All points $\vec{F}(x)$ where $\bar{x}_b \leq x \leq \bar{x}_i$ are on the Pareto frontier. Continue from S-9.

h. Case II:

$$f_1(x_i) > f_1(x_j)$$
$$f_2(x_i) < f_2(x_j)$$

Find the minimum $x_b < x_i$ where $f_1(x_b) = f_1(x_j)$. All points $\vec{F}(x)$ where $\bar{x}_b \leq x \leq \bar{x}_i$ are on the Pareto frontier. Continue from S-9.

S-10 Repeat from S-7 using the next non-dominated point

O-1 The Pareto frontier is identified in the steps of the algorithm.

Monte Carlo Method

I-1   Independent variables $\vec{x}$

           Objective function

I-2
$$\vec{F}(\vec{x}) = \begin{bmatrix} f_1(x) \\ f_2(x) \\ \vdots \\ f_n(x) \end{bmatrix}$$

I-3   Method for selecting points $\vec{x}$ randomly from the parameter space

S-1   Chose a point $\vec{x}$ from the parameter space

S-2   Compute the value of the objective function $\vec{F}(\vec{x})$

           Compare the value of $\vec{F}(\vec{x})$ with each $\vec{F}(\vec{p}_i)$ on the list of points on
           the frontier.

S-3
    i.   If $\vec{F}(\vec{x})$ dominates $\vec{F}(\vec{p}_i)$, remove $\vec{F}(\vec{p}_i)$ from the list of frontier points.
       Continue checking the frontier points.
    ii. If $\vec{F}(\vec{x})$ is dominated by $\vec{F}(\vec{p}_i)$, discard $\vec{F}(\vec{x})$ and continue from S-1.

S-4   If none of the points on the current list of frontier points dominates
           $\vec{F}(\vec{x})$, add $\vec{F}(\vec{x})$ to the list of frontier points.

S-5   Continue selecting points from S-1.

O-1   The list of frontier points $\vec{F}(\vec{p}_i)$ is the approximation to the Pareto
           frontier.

Weighted Sum Method

I-1   Independent variables $\vec{x}$

Objective function

I-2
$$\vec{F}(\vec{x}) = \begin{bmatrix} f_1(\vec{x}) \\ f_2(\vec{x}) \\ \vdots \\ f_n(\vec{x}) \end{bmatrix}$$

I-3   Method for optimizing a single objective function $g(\vec{x})$

Chose a point $\vec{\alpha}$ such that

S-1
$$\sum_{i=1}^{n} \alpha_i = 1 \quad |\alpha_i| \leq 1$$

Maximize the single objective function

S-2
$$g(\vec{x}) = \sum_{i=1}^{n} \alpha_i f_i(\vec{x})$$

Let $\vec{x}^*$ designate this maximum.

S-3   Compute the value of $\vec{F}(\vec{x}^*)$

S-4   Add $\vec{F}(\vec{x}^*)$ to the list of potential points on the Pareto frontier and continue from S-1 for the desired number of $\vec{\alpha}$ to compute

S-5   Iterate through the list of potential points for the Pareto frontier

S-6   Consider the ith point on the list $\vec{F}(\vec{x}_i^*)$

S-7   Compare $\vec{F}(\vec{x}_i^*)$ to each other point on the list of potential frontier points. If $\vec{F}(\vec{x}_i^*)$ is dominated by another point $\vec{F}(\vec{x}_j^*)$, eliminate $\vec{F}(\vec{x}_i^*)$ from the list. If $\vec{F}(\vec{x}_i^*)$ dominates any other point $\vec{F}(\vec{x}_k^*)$, eliminate $\vec{F}(\vec{x}_k^*)$ from the list.

O-1   The list of frontier points $\vec{F}(\vec{x}_i^*)$ remaining after S-7 is the approximation to the Pareto frontier.

Normal-Boundary Intersection

I-1   Independent variables $\vec{x}$

Objective function

I-2
$$F(\vec{x}) = \begin{bmatrix} f_1(\vec{x}) \\ f_2(\vec{x}) \\ \vdots \\ f_n(\vec{x}) \end{bmatrix}$$

I-3   Method for optimizing a single unconstrained objective function

I-4   Method for optimizing the single objective function $g(t) = t$ under a set of n constraints of the form $\Phi[\vec{w} + t\vec{u}] = \vec{F}(\vec{x})$

S-1   Compute the global optimum for each individual component of the objective function. Let $\vec{x}_i^*$ designate the global optimum for $f_i(\vec{x})$

Compute the matrix

S-2
$$\Phi = \begin{bmatrix} 0 & f_2(\vec{x}_2^*) - f_1(\vec{x}_2^*) & \cdots & f_n(\vec{x}_n^*) - f_1(\vec{x}_n^*) \\ f_1(\vec{x}_1^*) - f_2(\vec{x}_1^*) & 0 & \cdots & f_n(\vec{x}_n^*) - f_2(\vec{x}_n^*) \\ \vdots & \vdots & \vdots & \vdots \\ f_1(\vec{x}_1^*) - f_n(\vec{x}_1^*) & f_2(\vec{x}_2^*) - f_n(\vec{x}_2^*) & \cdots & 0 \end{bmatrix}$$

Choose a weight vector $\vec{w}$ such that

S-3
$$\sum_{i=1}^{n} w_i = 1 \quad |w_i| \le 1$$

Find the largest value t such that the n constraints

S-4
$$\Phi[\vec{w} + t\vec{u}] = \vec{F}(\vec{x})$$

are satisfied. Let $\vec{x}_i^b$ be the value of $\vec{x}$ corresponding to the maximum.

S-5   Save the point $\vec{F}(\vec{x}_i^b)$ as a candidate for the Pareto frontier.

S-6   Repeat from S-3 for as many points as desired.

S-7   Compare $\vec{F}(\vec{x}_i^b)$ to each other point on the list of potential frontier points. If $\vec{F}(\vec{x}_i^b)$ is dominated by another point $\vec{F}(\vec{x}_j^b)$, eliminate $\vec{F}(\vec{x}_i^b)$ from the list. If $\vec{F}(\vec{x}_i^b)$ dominates any other point $\vec{F}(\vec{x}_K^b)$, eliminate $\vec{F}(\vec{x}_i^b)$ from the list.

O-1   The list of frontier points $\vec{F}(\vec{x}_i^b)$ remaining after S-7 is the approximation to the Pareto frontier.

Normal Constraint

I-1   Independent variables $\vec{x}$

I-2   Objective function $\vec{F}(\vec{x})$ with components $f_n(\vec{x})$

I-3   Method for optimizing a single unconstrained objective function

I-4   Method for optimizing the single objective function $f_n(\vec{x})$ under a set of n constraints of the form $\vec{\vec{H}}^k \cdot \left(\vec{F}(\vec{x}) - \vec{X}^j\right) \geq 0$

S-1   Compute the global optimum for each individual component of the objective function. Let $\vec{x}_i^*$ designate the global optimum for $f_i(\vec{x})$

S-2   Compute the Nadir point $\vec{N} = \langle N_1, N_2, \ldots, N_n \rangle$ where
$$N_i = min[f_i(\vec{x}_1^*), f_i(\vec{x}_2^*), \ldots, f_i(\vec{x}_n^*)]$$

S-3   Transform the coordinates using
$$\bar{f}_i(\vec{x}) = \frac{f_i(\vec{x}) - N_i}{f_i(\vec{x}_i^*) - N_i}$$

S-4   Compute the $n-1$ hyperplane vectors that point from the point $(0, 0, \ldots, 0, 1)$ to each of the component maximums (these are unit vectors along each axis in the transformed coordinates)
$$\vec{\vec{H}}^k = \hat{e}^k - \hat{e}^n$$

S-5   Create a lattice of equally spaced points on the utopia hyperplane
$$\vec{\vec{X}}^j = \sum_{k=1}^{n} \alpha_{kj} \hat{e}^k \quad \sum_{k=1}^{n} \alpha_{kj} = 1 \quad |\alpha_{kj}| \leq 1$$

S-6   For each $\vec{\vec{X}}^i$, maximize $\bar{f}_n(\vec{x})$ subject to the constraints
$$\vec{\vec{H}}^k \cdot \left(\vec{F}(\vec{x}) - \vec{\vec{X}}^j\right) \geq 0$$
Let $\vec{x}_i^b$ be the corresponding point in parameter space.

S-7   Repeat from S-5 for each $\vec{\vec{X}}^i$.

S-8   Compare $\vec{F}(\vec{x}_i^b)$ to each other point on the list of potential frontier points. If $\vec{F}(\vec{x}_i^b)$ is dominated by another point $\vec{F}(\vec{x}_j^b)$, eliminate $\vec{F}(\vec{x}_i^b)$ from the list. If $\vec{F}(\vec{x}_i^b)$ dominates any other point $\vec{F}(\vec{x}_K^b)$, eliminate $\vec{F}(\vec{x}_i^b)$ from the list.

O-1   The list of frontier points $\vec{F}(\vec{x}_i^b)$ remaining after S-8 is the approximation to the Pareto frontier.

SPEA2

I-1	Independent variables $\vec{x}$
I-2	Objective function $\vec{F}(\vec{x})$ with components $f_n(\vec{x})$
I-3	Evolutionary method for generating a new population based on the fitness values for the states in the current population.
I-4	Size of the population pool N
I-5	Size of the archive $\overline{N}$
I-6	Size of the mating pool M
I-6	Value k for nearest neighbor distance metric. Commonly, $k = \sqrt{N + \overline{N}}$
S-1	Create a random initial population pool
S-2	Set the initial archive as an empty set
S-3	Create the union set by combining together the states of the archive and population pool. The union set and archive set contain only unique members, whereas the population pool may contain duplicates.
S-4	For each state i in the union count the number of states dominated by i. Let p_i be the number of states dominated by state i.

S-5

For each state i, compute

$$S_i = \sum_{i \prec j} p_j$$

Given a state i, find all states j that dominate state i $(i \prec j)$. Sum the value of p_j for each state j that dominates state i.

S-6

For each state i in the union count the distance to every other state in the union. Sort the distances to the other states by increasing order of distance. Let d_{ij} be the distance from state i to the jth state in rank order with smaller distances associated with smaller values of j.

S-7

Compute the distance metric

$$D_i = \frac{1}{d_{ik} + 2}$$

where k is set in I-6.

S-8 Compute the fitness for the ith state as

$$F_i = S_i + D_i$$

S-9 Determine all states in the union that are not dominated by any other state. These states be identified from the previous computations in S-5 as the states where $S_i = 0$.

S-10 Replace the archive set with the set of unique states in the union that are not dominated.

2. If the resulting number of states in the archive set is exactly $\overline{N}$, then continue to S-11.

3. If the resulting number of states in the archive set is smaller than $\overline{N}$, then add in the states from the union set of the previous generation until the current archive has $\overline{N}$ states.

4. If the resulting number of states in the archive set is larger than $\overline{N}$, then remove states from the archive by:

    a. Identify the state with the smallest value of d_{i1} (distance to its nearest neighbor). If multiple states are tied for the same smallest value, break the tie by considering the d_{i2}, etc.

    b. Remove from the archive the state identified in substep (3)(a) above.

S-11 Perform binary tournament selection with replacement on the new archive set to create a mating pool.

2. Clear the previous mating pool.

3. Choose two states from the archive set.

4. Copy the state with the *lower* fitness to the mating pool.

5. Do not remove these states from the archive set.

6. Repeat from substep (2) above until the population pool has M states.

S-12 Clear the population pool and apply recombination / mutation on the mating pool to create a new population pool.

S-13 Repeat from S-3 as many times as desired or until sufficient nondominated points are identified.

O-1 The list of states in the archive set is the approximation to the Pareto frontier.

NSGA-II

I-1	Independent variables $\vec{x}$
I-2	Objective function $\vec{F}(\vec{x})$ with components $f_n(\vec{x})$
I-3	Evolutionary method for generating a new population based on the fitness values for the states in the current population.
I-4	Size of the population pool N
S-1	Create a random initial population pool of N parent states $P_{(0)}$
S-2	Create an initial population pool of N child states $C_{(0)}$ from $P_{(0)}$ by using binary tournament selection with replacement, crossover, and mutation
S-3	Create the set $T = P_{(n)} \cup C_{(n)}$

For each state $T_k \in T$:

S-4

    5. Count the number of states in T that dominate T_k (non-dominance level)

    6. Create a list of all states T_s where T_k dominates T_s (dominance list)

For each state $T_k \in T$ where the non-dominance level is zero:

    7. Remove the state T_k

    8. Add T_k to the set $\mathcal{F}_1$

S-5

    9. For each state T_l on the dominance list for T_k, subtract one from the non-dominance level of T_l.

Repeat these steps again, but in step (2) assign the states to the set $\mathcal{F}_2$. Continue repeating this process until T is empty. For each iteration, assign the states to a new set $\mathcal{F}_n$.

Set $P_{n+1} = \mathcal{F}_1$. If $|P_{n+1}| \geq N$ continue to S-7.

S-6   Otherwise set $P_{n+1} = \mathcal{F}_1 \cup \mathcal{F}_2$. If $|P_{n+1}| \geq N$ continue to S-7.

Otherwise, continue to add additional sets $\mathcal{F}_k$ until $|P_{n+1}| \geq N$.

Let k_{max} be the maximum value for $\mathcal{F}_k$ required in S-6. For each $\mathcal{F}_k$ where $k = 1 \dots k_{max}$ compute the crowding distance:

S-7

    4. Initialize the crowding distance for each state in $\mathcal{F}_k$ to zero

    5. For each component i of the objective function

        a. Sort the states in $\mathcal{F}_k$ by the value of component i

        b. Let $\mathcal{F}_k^{(i)}$ represent the sequence formed from sorting $\mathcal{F}_k$ by the ith component

    c. Let $\mathcal{F}_k^{(i)(j)}$ be the jth state in the ordered sequence where $j = 1$ is the lowest value of the component and the sequence increases in value

    d. Let $\mathcal{F}_k^{(i)(j)}{}_n$ be the value of the nth component of the of the jth state in the sorted sequence $\mathcal{F}_k^{(i)}$

    e. For the first and last state in the sequence $\mathcal{F}_k^{(i)}$, set the crowding distance to infinity

    f. For every other state in the sequence $\mathcal{F}_k^{(i)}$, increment the crowding distance for the state by the value

$$\mathcal{F}_k^{(i)(j+1)}{}_i - \mathcal{F}_k^{(i)(j-1)}{}_i$$

    g. Repeat these steps for each component of $\mathcal{F}_k$

  6. Repeat these steps for each $\mathcal{F}_k$

S-8

Sort P_{n+1} using the total ordering operator

$$S_i <_n S_j \rightarrow \left(ND_i < ND_j\right) \vee \left[\left(ND_i = ND_j\right) \wedge \left(C_i > C_j\right)\right]$$

where $S_i, S_j \in P_{n+1}$, ND_i is the non-dominance level for S_i as computed in S-5, and C_i is the crowding distance as computed in S-7. Let $P_{n+1}^{(i)}$ be the sorted sequence where $i = 1$ corresponds to the best values (ND_i is lowest and C_i is highest).

S-9

Truncate $P_{n+1}^{(i)}$ to the first N states in the sequence

S-10

Perform binary tournament selection with replacement on the new archive set to create the next generation child set $C_{(n+1)}$.

  1. Choose one or more states from $P_{n+1}^{(i)}$

  2. Use selection, recombination, mutation to generate a new state based on the states selected

  3. Do not remove these states from the archive set

  4. Add the new state to $C_{(n+1)}$

  5. Continue from (1) until $\left|C_{(n+1)}\right| = N$

S-11

Repeat from S-3 as many times as desired or until sufficient nondominated points are identified.

O-1

The list of nondominated states in $T = P_{(n+1)} \cup C_{(n+1)}$ in the archive set is the approximation to the Pareto frontier.

Directed Search Domain

I-1   Independent variables $\vec{x}$

I-2   Objective function $\vec{F}(\vec{x})$ with components $f_n(\vec{x})$

I-3   Evolutionary method for generating a new population based on the fitness values for the states in the current population.

I-4   Number of hyperplane points to consider N

I-5   Method for selecting a random angle γ on the range $\left[0, \frac{\pi}{4}\right)$

I-6   Method for selecting a random directional vector $\hat{c}$

I-7   Method for optimizing a single unconstrained objective function

S-1   Compute the global optimum for each individual component of the objective function. Let $\vec{x}_i^*$ designate the global optimum for $f_i(\vec{x})$

S-2   Compute the Nadir point $\vec{N} = \langle N_1, N_2, ..., N_n \rangle$ where
$$N_i = min[f_i(\vec{x}_1^*), f_i(\vec{x}_2^*), ..., f_i(\vec{x}_n^*)]$$

S-3   Transform the coordinates using
$$\bar{f}_i(\vec{x}) = \frac{f_i(\vec{x}) - N_i}{f_i(\vec{x}_i^*) - N_i}$$

S-4   Create a lattice of equally spaced points on the utopia hyperplane
$$\vec{\bar{X}}^j = \sum_{k=1}^{n} \alpha_{kj}\hat{e}^k \qquad \sum_{k=1}^{n} \alpha_{kj} = 1 \quad |\alpha_{kj}| \leq 1$$

S-5   For every $\vec{\bar{X}}^j$, optimize the single objective function
$$max \sum_{i=1}^{n} \vec{F}_i(\vec{x})$$
subject to the constraint
$$\frac{\sum_{i=1}^{n}\left(F_i(\vec{x}) - \alpha_{ji}\right)c_i}{\sqrt{\sum_{i=1}^{n}\left(F_i(\vec{x}) - \alpha_{ji}\right)^2}} \geq cos\,\gamma$$
where γ is a random angle on the range $\left[0, \frac{\pi}{4}\right)$ and $\hat{c}$ is a directional vector in the direction of the axis of the constraint cone.

O-1   The list of nondominated states in $T = P_{(n+1)} \cup C_{(n+1)}$ in the archive set is the approximation to the Pareto frontier.

Random Weighted Genetic Algorithm

I-1   Independent variables $\vec{x}$

I-2   Objective function

$$F(\vec{x}) = \begin{bmatrix} f_1(\vec{x}) \\ f_2(\vec{x}) \\ \vdots \\ f_n(\vec{x}) \end{bmatrix}$$

I-3   Method for choosing a random weight vector $\vec{\alpha}$ where

$$\sum_{i=1}^{n} \alpha_i = 1 \quad |\alpha_i| \leq 1$$

I-4   Population size N

I-5   Population to remove and replace from the archive n_e

S-1   Create an initial population of states chosen at random

S-2   Generate a random weight vector $\vec{\alpha}$

S-3   Compute the fitness value for each state as

$$f = \sum_{i=1}^{n} \alpha_i F_i(\vec{x})$$

S-4   Compute the minimum fitness across all states f_{min}

S-5   Compute the selection probability for each state as

$$p_i = \frac{f_i - f_{min}}{\sum_{k=0}^{N} [f_k - f_{min}]}$$

S-6   Select states according to the selection probability, apply crossover and mutation to create N child states

S-7   Add the child states to the archive and retain all nondominated solutions in the archive

S-8   Remove n_E states at random from the child population. Replace these with randomly selected states from the archive

O-1   The archive represents the approximation to the Pareto frontier

Utopia Lattice Weighted Sum

I-1   Independent variables $\vec{x}$

       Objective function

I-2
$$\vec{F}(\vec{x}) = \begin{bmatrix} f_1(\vec{x}) \\ f_2(\vec{x}) \\ \vdots \\ f_n(\vec{x}) \end{bmatrix}$$

I-5   Population size N

S-1   Compute the global optimum for each individual component of the objective function. Let $\vec{x}_i^*$ designate the global optimum for $f_i(\vec{x})$

S-2   Compute the Nadir point $\vec{N} = \langle N_1, N_2, \dots, N_n \rangle$ where
$$N_i = min[f_i(\vec{x}_1^*), f_i(\vec{x}_2^*), \dots, f_i(\vec{x}_n^*)]$$

S-3   Transform the coordinates using
$$\bar{f}_i(\vec{x}) = \frac{f_i(\vec{x}) - N_i}{f_i(\vec{x}_i^*) - N_i}$$

S-4   Create a lattice of equally spaced points on the utopia hyperplane
$$\vec{X}^j = \sum_{k=1}^{n} \alpha_{kj}(\hat{\varepsilon}_i - \hat{\varepsilon}_n) \quad \sum_{k=1}^{n} \alpha_{kj} = 1 \quad |\alpha_{kj}| \leq 1$$

S-5   For each value j, maximize the single objective function
$$g(\vec{x}) = \sum_{i=1}^{n} \alpha_{ij} f_i(\vec{x})$$

       Let $\vec{x}_j^*$ designate this maximum.

S-6   Compute the value of $\vec{F}(\vec{x}_j^*)$

S-7   Add $\vec{F}(\vec{x}_j^*)$ to the archive and retain all nondominated solutions in the archive

O-1   The archive represents the approximation to the Pareto frontier

Vector Evaluated Genetic Algorithm

I-1   Independent variables $\vec{x}$

Objective function

I-2
$$\vec{F}(\vec{x}) = \begin{bmatrix} f_1(\vec{x}) \\ f_2(\vec{x}) \\ \vdots \\ f_n(\vec{x}) \end{bmatrix}$$

I-3   Population size N, where $N \bmod n \equiv 0$

S-1   Create an initial population of N states chosen at random

S-2   Create a sequence from the population where the sequence order is random

S-3   Divide the population into n equal groups

S-4   Set the fitness for the states in the kth group equal to the value of $F_k(\vec{x})$ for the state

S-5   Compute the minimum fitness across all states f_{min}

Compute the selection probability for each state as

S-6
$$p_i = \frac{f_i - f_{min}}{\sum_{k=0}^{N}[f_k - f_{min}]}$$

S-7   Use selection, recombination, and mutation to create a new population

S-8   Add the new states to an archive maintaining only nondominated states

O-1   The archive represents the approximation to the Pareto frontier

Pareto Envelope Based Selection Algorithm II

I-1 Independent variables $\vec{x}$

I-2 Objective function

$$\vec{F}(\vec{x}) = \begin{bmatrix} f_1(\vec{x}) \\ f_2(\vec{x}) \\ \vdots \\ f_n(\vec{x}) \end{bmatrix}$$

I-3 Population size N

I-4 Number of segments s for dividing each component

S-1 Create an initial population of N states chosen at random

S-2 Divide the operational space into n^s cells, where n is the number of components for the objective function

S-3 Count the number of states in each cell

S-4 Set the fitness of a cell equal to the number of states in the cell

S-5 Discard all cells that have no states

S-6 Use binary tournament selection to select cells in proportion to their fitness

S-7 From a selected cell, randomly choose a state from the cell

S-8 Use recombination and mutation on the chosen state(s) to create a new state

S-9 Repeat from S-6 to create a new population of N states

S-10 Add the new states to the archive of nondominated states. Update the archive so that the set of states in the archive are all nondominated.

S-11 Set the population to the new population created from cell selection above

S-12 Repeat from S-3 for each iteration

O-1 The archive represents the approximation to the Pareto frontier

Dynamic Multiobjective Evolutionary Algorithm

I-1   Independent variables $\vec{x}$

I-2   Objective function

$$F(\vec{x}) = \begin{bmatrix} f_1(\vec{x}) \\ f_2(\vec{x}) \\ \vdots \\ f_n(\vec{x}) \end{bmatrix}$$

I-3   Population size N

S-1   Create an initial population of N states chosen at random. Designate the population of states as $\mathcal{T}$.

S-2   Assign each state a nondominance rank equal to the number of states that dominate the state. Set R_i as the nondominance rank for the ith state.

S-3   Compute the crowding distance C_i for each state:

4.   Initialize the crowding distance for each state to zero
5.   For each component i of the objective function
   a.   Sort the states in by the value of component i
   b.   Let $\mathcal{T}^{(i)}$ represent the sequence formed from sorting $\mathcal{T}$ by the ith component
   c.   Let $\mathcal{T}^{(i)(j)}$ be the jth state in the ordered sequence where $j = 1$ is the lowest value of the component and the sequence increases in value
   d.   Let $\mathcal{T}^{(i)(j)}{}_n$ be the value of the nth component of the of the jth state in the sorted sequence $\mathcal{T}^{(i)}$
   e.   For the first and last state in the sequence $\mathcal{T}^{(i)}$, set the crowding distance to infinity
   f.   For every other state in the sequence $\mathcal{T}^{(i)}$, increment the crowding distance for the state by the value

   $$\mathcal{T}^{(i)(j+1)}{}_i - \mathcal{T}^{(i)(j-1)}{}_i$$

   g.   Repeat these steps for each component of $\mathcal{T}$
6.   Repeat these steps for each $\mathcal{T}$

S-4   Compute the entropy for each state as

$$S_i = -p_i \log p_i$$

where

$$p_i = \frac{e^{-R_i/T}}{\sum_{k=1}^{n} e^{-R_k/T}}$$

S-5

Compute the fitness for each state as

$$F_i = R_i - TS_i - C_i$$

S-6

Use binary tournament selection to create a child population based on the fitness of the population $\mathcal{T}$

S-7

Compute the metrics from S-2, S-3, and S-4 for each of the child states.

S-8

Compute the fitness for each of the child states.

S-9

Each of the child states is selected or rejected by comparing the metrics of the child state to the metrics of the current worst fitness of the parent population. A child is selected if any of the following conditions are met:

7. The nondominance rank of the child is less than the nondominance rank of the worst parent
8. If the nondominance rank of the child is equal to the nondominance rank of the worst parent, and the crowding distance of the child is greater than the crowding distance of the parent
9. If

$$exp\left[(R_{worst} - R_{daughter})/T\right] > r$$

where r is a random number generated uniformly on the range $[0,1)$

If a child is selected, the worst parent is removed from the population and the child is added to the population and the new worst state is identified. The process is repeated for each of the child states.

S-10

Add the child population to the archive of nondominated states. Update the archive so that the set of states in the archive are all nondominated.

S-11

Set the population to the child population created

O-1

The archive represents the approximation to the Pareto frontier

Pareto Archive Evolution Strategy

I-1　　　Independent variables $\vec{x}$

　　　　Objective function

I-2

$$F(\vec{x}) = \begin{bmatrix} f_1(\vec{x}) \\ f_2(\vec{x}) \\ \vdots \\ f_n(\vec{x}) \end{bmatrix}$$

I-3　　　Population size N

S-1　　　Create an initial population of N states chosen at random

S-2　　　For each state, apply the mutation operator to create a new state.

S-3　　　If the new state is dominated by the original state, reject the new state and continue from S-2 for the next state

S-4　　　If the new state dominates the original state, then replace the original state with the new state, add the new state to the archive, and continue from S-2 for the next state

S-5　　　If neither state dominates the other, then compare the new state with the archive. If the new state is not dominated by any state in the archive, then replace the original state with the new state. If the new state is dominated by any state in the archive, reject the new state.

S-6　　　Continue from S-2 for the next state.

S-7　　　Repeat the entire process from S-2 for each iteration of the algorithm

O-1　　　The archive represents the approximation to the Pareto frontier

Niched Pareto Genetic Algorithm

I-1   Independent variables $\vec{x}$

Objective function

I-2
$$\vec{F}(\vec{x}) = \begin{bmatrix} f_1(\vec{x}) \\ f_2(\vec{x}) \\ \vdots \\ f_n(\vec{x}) \end{bmatrix}$$

I-3   Population size N

I-4   Minimum sharing distance σ_{min}

S-1   Create an initial population of N states chosen at random

S-2   Set the fitness of each state to zero

For each state, count the number of states where $d < \sigma_{min}$. For each state where this condition add a factor of

S-3
$$S = 1 - \frac{d}{\sigma_{min}}$$

to the fitness for the state.

S-4   Apply tournament selection with replacement, mutation, and recombination to generate a new population based on the current population

S-5   Add the new population to the archive of nondominated states. Update the archive to maintain nondominance among the states.

S-6   Set the population to the new population

S-7   Repeat from S-2 for as many iterations as desired

O-1   The archive represents the approximation to the Pareto frontier

Weighted Metric Method

I-1 Independent variables $\vec{x}$

Objective function

I-2
$$\vec{F}(\vec{x}) = \begin{bmatrix} f_1(\vec{x}) \\ f_2(\vec{x}) \\ \vdots \\ f_n(\vec{x}) \end{bmatrix}$$

I-3 Method for optimizing a single objective function $g(\vec{x})$

I-4 Reference point $\vec{z}$

I-5 Value of p for the metric

Chose a point $\vec{w}$ such that

S-1
$$\sum_{i=1}^{n} w_i = 1 \quad |w| \le 1$$

Maximize the single objective function

S-2
$$g(\vec{x}) = \left(\sum_{i=1}^{n} w_i [f_i(\vec{x}) - z_i]^p \right)^{1/p}$$

Let $\vec{x}^*$ designate this maximum.

S-3 Compute the value of $\vec{F}(\vec{x}^*)$

S-4 Add $\vec{F}(\vec{x}^*)$ to the list of potential points on the Pareto frontier and continue from S-1 for the desired number of $\vec{w}$ to compute

S-5 Iterate through the list of potential points for the Pareto frontier

S-6 Consider the ith point on the list $\vec{F}(\vec{x}_i^*)$

S-7 Compare $\vec{F}(\vec{x}_i^*)$ to each other point on the list of potential frontier points. If $\vec{F}(\vec{x}_i^*)$ is dominated by another point $\vec{F}(\vec{x}_j^*)$, eliminate $\vec{F}(\vec{x}_i^*)$ from the list. If $\vec{F}(\vec{x}_i^*)$ dominates any other point $\vec{F}(\vec{x}_k^*)$, eliminate $\vec{F}(\vec{x}_k^*)$ from the list.

O-1 The list of frontier points $\vec{F}(\vec{x}_i^*)$ remaining after S-7 is the approximation to the Pareto frontier.

ε-Constraint Method

I-1 Independent variables $\vec{x}$

Objective function

I-2
$$\vec{F}(\vec{x}) = \begin{bmatrix} f_1(\vec{x}) \\ f_2(\vec{x}) \\ \vdots \\ f_n(\vec{x}) \end{bmatrix}$$

I-3 Method for optimizing a single objective function $g(\vec{x})$

I-4 Component to optimize c

S-1 Use the single objective method to maximize each component of the objective function individually. Set μ_i as the maximum component value for the ith component.

S-2 Use the single objective method to minimize each component of the objective function individually. Set ν_i as the minimum component value for the ith component.

S-3 Choose a set of parameters ε_i where $\nu_i \leq \varepsilon_i \leq \mu_i$

Maximize the single objective function

$$g(\vec{x}) = f_c(\vec{x})$$

S-4 subject to the constraints

$$f_i(\vec{x}) \geq \varepsilon_i \quad i \neq c$$

Let $\vec{x}^*$ designate this maximum.

S-5 Compute the value of $\vec{F}(\vec{x}^*)$

S-6 Compare $\vec{F}(\vec{x}_i^*)$ to each other point on the archive of potential frontier points and adjust the archive to maintain a list of nondominated states

S-7 Repeat from S-3 as many times as desired

O-1 The archive is the approximation to the Pareto frontier

Keeney-Raiffa Method

I-1   Independent variables $\vec{x}$

Objective function

I-2
$$\vec{F}(\vec{x}) = \begin{bmatrix} f_1(\vec{x}) \\ f_2(\vec{x}) \\ \vdots \\ f_n(\vec{x}) \end{bmatrix}$$

I-3   Method for optimizing a single objective function $g(\vec{x})$

Chose a point $\vec{w}$ such that

S-1
$$\sum_{i=1}^{n} w_i = 1 \quad |w| \leq 1$$

Maximize the single objective function

S-2
$$g(\vec{x}) = \prod_{k=1}^{n} [1 + w_i f_i(\vec{x})]$$

Let $\vec{x}^*$ designate this maximum.

S-3   Compute the value of $\vec{F}(\vec{x}^*)$

S-4   Compare $\vec{F}(\vec{x}_i^*)$ to each other point on the list of potential frontier points. If $\vec{F}(\vec{x}_i^*)$ is dominated by another point $\vec{F}(\vec{x}_j^*)$, eliminate $\vec{F}(\vec{x}_i^*)$ from the list. If $\vec{F}(\vec{x}_i^*)$ dominates any other point $\vec{F}(\vec{x}_k^*)$, eliminate $\vec{F}(\vec{x}_k^*)$ from the list.

O-1   The list of frontier points $\vec{F}(\vec{x}_i^*)$ remaining after S-7 is the approximation to the Pareto frontier.

Appendix B: Gaussian Sampling

Some of the optimization algorithms discussed requires generating random numbers for a Gaussian distribution. In this appendix we provide methods for generating random variables from single and multivariate Gaussian distributions.

Single Variate Gaussian Sampling

In the case of a single Gaussian distributed variable, the probability density function is

$$p(x) = \frac{1}{\sqrt{2\pi\sigma}} exp\left(-\frac{(x-\mu)^2}{2\sigma}\right) \tag{B.1}$$

where μ is the distribution mean and σ^2 is the variance.

In most cases there is an existing capability to generate uniform random deviates on the range $[0,1)$. Let U_1 and U_2 represent two such uniform random varaibles. Then by the Box-Muller transform, the varaibles

$$Z_1 = \sqrt{-2 \ln U_1} \cos(2\pi U_2) \tag{B.2}$$

$$Z_2 = \sqrt{-2 \ln U_1} \sin(2\pi U_2) \tag{B.3}$$

are independent random Gaussian distributed variables with mean zero and variance 1. These variables may be transformed using

$$\bar{Z}_1 = \mu + \sigma Z_1 \tag{B.4}$$

$$\bar{Z}_2 = \mu + \sigma Z_2 \tag{B.5}$$

where the bar-variables are independent Gaussian distributed random variables with mean μ and variance σ^2.

The Box-Muller method creates two independent Gaussian random variables from two independent uniform variables. It is important to keep in mind that even though these two variables are created together, they are actually independent of each other in terms of their distributions. This method may also be used when generating multinomial Gaussian variables as shown in the section below.

Multinomial Gaussian Sampling

The probability density function for the multinomial distribution is given by

$$p(\vec{x}) = \frac{1}{(2\pi)^{k/2}\sqrt{|\Sigma|}} exp\left(-\frac{1}{2}(\vec{x}-\vec{\mu})^T \Sigma^{-1}(\vec{x}-\vec{\mu})\right) \qquad \text{B.6}$$

where $\vec{\mu}$ is the mean of the distribution and Σ is the covariance matrix

$$\Sigma = \begin{bmatrix} \langle(x_1-\mu_1)(x_1-\mu_1)\rangle & \langle(x_1-\mu_1)(x_2-\mu_2)\rangle & \cdots & \langle(x_1-\mu_1)(x_n-\mu_n)\rangle \\ \langle(x_2-\mu_2)(x_1-\mu_1)\rangle & \langle(x_2-\mu_2)(x_2-\mu_2)\rangle & \cdots & \langle(x_2-\mu_2)(x_n-\mu_n)\rangle \\ \vdots & \vdots & \vdots & \vdots \\ \langle(x_n-\mu_n)(x_1-\mu_1)\rangle & \langle(x_n-\mu_n)(x_2-\mu_2)\rangle & \cdots & \langle(x_n-\mu_n)(x_n-\mu_n)\rangle \end{bmatrix} \qquad \text{B.7}$$

or

$$\Sigma = \begin{bmatrix} \sigma_{11} & \sigma_{12} & \cdots & \sigma_{1n} \\ \sigma_{21} & \sigma_{22} & \cdots & \sigma_{2n} \\ \vdots & \vdots & \vdots & \vdots \\ \sigma_{n1} & \sigma_{n2} & \cdots & \sigma_{nn} \end{bmatrix} \qquad \text{B.8}$$

where

$$\sigma_{ij} = \langle(x_i-\mu_i)(x_j-\mu_j)\rangle \qquad \text{B.9}$$

Random variables conforming to this distribution may be generated as follows. First, identify any matrix A such that

$$AA^T = \Sigma \qquad \text{B.10}$$

Next, generate N independent Gaussian distributed random variables with mean zero and variance 1. Here, N is the number of variables in the distribution. Set

$$\vec{z} = [z_1, z_2, \dots, z_N] \qquad \text{B.11}$$

where z_i are the Gaussian distributed varaibles.

The vector

$$\vec{r} = \vec{\mu} + A\vec{z} \qquad \text{B.12}$$

is vector with N components where the components of the vector are distributed according to the multinomial Gaussian distribution with mean $\vec{\mu}$ is and covariance matrix Σ.

Appendix C: Multiobjective Optimization Concepts

C.1 Pareto Comparison

Let $\vec{v}^1, \vec{v}^2$ be two vectors in a n-dimensional space where $n \geq 2$. Let $<_n$ be the map $<_n : \mathbb{R}^n \times \mathbb{R}^n \to \{0,1\}$ defined as

$$\vec{v}^1 <_n \vec{v}^2 \to \left(\vec{v}_i^1 \leq \vec{v}_i^2 \ \forall \ i = 1 \dots n \right) \wedge \left(\exists k \mid \vec{v}_k^1 < \vec{v}_k^2 \right) \qquad \text{C.1}$$

In words, $\vec{v}^1 <_n \vec{v}^2$ means that the value of each component of $\vec{v}^1$ is less than or equal to the corresponding value of $\vec{v}^2$, and there exist at least one component that is strictly less. The second condition assures that $\vec{v}^1 \not<_n \vec{v}^1$. Without this condition, a vector would be less than itself.

The comparison operator above extends the concept of 'less than' to vector values in two or higher dimensions. In one dimension, given two numbers x and y, one of three conditions must be true: $x < y$, $y < x$, or $x = y$. In fact, we can reduce this to just two possibilities by considering 'less than or equal to': $x \leq y$ or $y \leq x$, or both.

When a comparison must result in one of these options possibilities, the operator $\leq$ creates a total ordering on the set. This means that given any two elements in the set it is always possible to order the elements relative to each other.

The 'less than' extension provided above is a little different. Given two vectors $\vec{v}^1$ and $\vec{v}^2$, it may be true that $\vec{v}^1 <_n \vec{v}^2$, $\vec{v}^2 <_n \vec{v}^1$, $\vec{v}^1 = \vec{v}^2$, or none of the three. This operator can be extended to 'less than or equal to' as well:

$$\vec{v}^1 \leq_n \vec{v}^2 \to \left(\vec{v}_i^1 \leq \vec{v}_i^2 \ \forall \ i = 1 \dots n \right) \qquad \text{C.2}$$

This is similar to the $<_n$ but does not include the second condition. Given two vectors $\vec{v}^1$ and $\vec{v}^2$ we can have $\vec{v}^1 \leq_n \vec{v}^2$, $\vec{v}^2 \leq_n \vec{v}^1$, $\vec{v}^1 = \vec{v}^2$, or none of the three.

For example, consider the vectors $\vec{v}^1 = (1,0)$ and $\vec{v}^2 = (0,1)$. From the definition of the $<_n$, we need to examine each component of the vectors. We have

$$\vec{v}_1^2 < \vec{v}_1^1 \qquad \text{C.3}$$

$$\vec{v}_2^2 \not< \vec{v}_2^1 \qquad \text{C.4}$$

The first condition in C.1 is not met so $\vec{v}^2 \not\prec_n \vec{v}^1$. Alternatively,

$$\vec{v}_1^1 \not\prec \vec{v}_1^2 \qquad\qquad \text{C.5}$$

$$\vec{v}_2^1 < \vec{v}_2^2 \qquad\qquad \text{C.6}$$

Again, the first condition in C.1 is not met so $\vec{v}^1 \not\prec_n \vec{v}^2$. Finally,

$$\vec{v}_1^1 \neq \vec{v}_1^2 \qquad\qquad \text{C.7}$$

$$\vec{v}_2^1 \neq \vec{v}_2^2 \qquad\qquad \text{C.8}$$

so $\vec{v}^1 \neq \vec{v}^2$. In this case none of the three options is true.

The comparison operator $\leq_n$ is a partial ordering operation that may be used to define a poset. A partial ordering operation has the properties

reflexivity $\qquad\qquad a \leqslant a \qquad\qquad$ C.9

antisymmetry $\qquad a \leqslant b \wedge b \leqslant a \Rightarrow a \asymp b \qquad$ C.10

transitivity $\qquad a \leqslant b \wedge b \leqslant c \Rightarrow a \leqslant c \qquad$ C.11

where $\leqslant$ is analogous to $\leq$, $\asymp$ is analogous to $=$, $\wedge$ is a logical 'and', and $\Rightarrow$ is the logical implication (if a then b). Each of these conditions is met by the comparison operator $\leq_n$.

C.2 Multiobjective Optimization Definition

The general multiobjective optimization problem can be written as

$$max\ \vec{F}(\vec{x}) = \begin{bmatrix} f_1(\vec{x}) \\ f_2(\vec{x}) \\ \vdots \\ f_n(\vec{x}) \end{bmatrix}$$
$$g_1(\vec{x}) = 0$$
$$g_2(\vec{x}) = 0 \qquad\qquad \text{C.12}$$
$$\vdots$$
$$g_k(\vec{x}) = 0$$
$$h_1(\vec{x}) \leq 0$$
$$h_2(\vec{x}) \leq 0$$
$$\vdots$$
$$h_l(\vec{x}) \leq 0$$

The desire is to simultaneously maximize all components of the objective function $\vec{F}(\vec{x})$ subject to the k equality constraints $g_k(\vec{x}) = 0$ and the l inequality constraints $h_l(\vec{x}) \leq 0$.

C.3   Pareto Frontier

Given a multiobjective optimization problem in the form of C.12 and the comparison operator C.1, if there is at least one point $\vec{x}$ satisfying the constraint equations $g(\vec{x}) = 0$ and $h(\vec{x}) \leq 0$, then there is at least one point $\vec{F}(\vec{x}_i)$ such that there is no other point $\vec{x}_j$ where $\vec{F}(\vec{x}_i) <_n \vec{F}(\vec{x}_j)$.

A point $\vec{F}(\vec{x}_i)$ is said to be dominated by the point $\vec{F}(\vec{x}_j)$ if $\vec{F}(\vec{x}_i) <_n \vec{F}(\vec{x}_j)$. If two points are related such that $\vec{F}(\vec{x}_i) \not<_n \vec{F}(\vec{x}_j)$ and $\vec{F}(\vec{x}_j) \not<_n \vec{F}(\vec{x}_i)$ then the points $\vec{F}(\vec{x}_i)$ and $\vec{F}(\vec{x}_j)$ are called incomparable.

The Pareto frontier is the set of points $\vec{F}(\vec{x}_i)$ where $\vec{x}_i$ satisfies the constraints $g(\vec{x}_i) = 0$ and $h(\vec{x}_i) \leq 0$, and where there is no other point $\vec{F}(\vec{x}_j)$ where $\vec{F}(\vec{x}_i) <_n \vec{F}(\vec{x}_j)$ when $\vec{x}_j$ also satisfies the constraints $g(\vec{x}_j) = 0$ and $h(\vec{x}_j) \leq 0$.

The Pareto frontier is the set of points that are not dominated by any other point. These points represent the potential optima for the multiobjective optimization problem from C.12.

C.4   Utopia Hyperplane

Consider a multiobjective optimization problem of C.12. Optimize each component of the objective function independently. Let the vector $\vec{x}_i^*$ be the point in parameter space that maximizes $f_i(\vec{x})$ subject to the constraints $g(\vec{x}) = 0$ and $h(\vec{x}) \leq 0$.

Let $\vec{F}_i^* = \vec{F}(\vec{x}_i^*)$. This is the point in the operational space where the ith component has its maximum value. The set of vectors $\vec{F}_i^*$ for $i = 1 \ldots n$ are the utopia vectors for the optimization problem. These vectors point to the maximum values for each individual component of the objective.

Each of the anchor points must be on the Pareto frontier. Since these points have the maximum value for at least one component of the objective function, there can be no point that dominates the anchor point.

This definition of the utopia hyperplane assumes that each component has a unique point $\vec{x}_i^*$ that maximizes the value of the component. This condition is not necessarily true. For two dimensional problems, if there are multiple points $\vec{x}_i^*$ that maximize the component, we chose the point with the largest value of the other component.

However, in higher dimensions, if there are multiple $\vec{x}_i^*$ that maximize one component, the remaining components create a separate multiobjective and there may not be any dominant point. In this case, the utopia hyperplane is the hyperplane that divides the operational space into two components where all of the anchor points (including the multiple $\vec{x}_i^*$) lie either on the hyperplane or on the same side, and where the hyperplane contains the maximum number of anchor points from different components.

C.5   Utopia Point

The utopia point is the point constructed from the maximum value for each component of the objective function. Let $\vec{x}_i^*$ be the point that maximizes the ith component of the objective function:

$$\vec{F}(\vec{x}) = \begin{bmatrix} f_1(\vec{x}_1^*) \\ f_2(\vec{x}_2^*) \\ \vdots \\ f_n(\vec{x}_n^*) \end{bmatrix} \qquad \text{C.13}$$

The utopia point is typically not an allowable point in the objective space. If the utopia point is a realizable point in the objective space, then this point must be the unique optimum for the multiobjective point because this point dominates all other points in the objective space.

C.6   Nadir Point

The Nadir point is computed from the anchor points. Given a set of anchor points $\vec{x}_i^*$, the Nadir point is given by

$$\vec{N} = \langle N_1, N_2, ..., N_n \rangle \qquad \text{C.14}$$

where

$$N_i = min[f_i(\vec{x}_1^*), f_i(\vec{x}_2^*), ..., f_i(\vec{x}_n^*)] \qquad \text{C.15}$$

Once we have the set of n points $\vec{x}_i^*$ (one for each component of the objective function), we compute the minimum value for each component across all of the anchor points. We then combine this together for each of the components of the objective function.

C.7   Miettinen Transformation

The Miettinen transformation independently transforms each component of the objective function according to the components of the Nadir point:

$$\bar{f}_i(\vec{x}) = \frac{f_i(\vec{x}) - N_i}{f_i(\vec{x}_i^*) - N_i} \qquad \text{C.16}$$

With respect to the points $\vec{x}_i^*$, the transformed components are all on the range [0,1].

www.ingramcontent.com/pod-product-compliance
Lightning Source LLC
Chambersburg PA
CBHW060324200326
41519CB00011BA/1832